## Primary SPACE Project Research Team


### Research Co-ordinating Group

Professor Paul Black (Co-director)
Jonathan Osborne

Centre for Educational Studies
King's College London
University of London
Cornwall House Annexe
Waterloo Road
London SE1 8TZ

Tel: 071 872 3094

Dr Wynne Harlen (Co-director)
Terry Russell

Centre for Research in Primary Science
and Technology
Department of Education
University of Liverpool
126 Mount Pleasant
Liverpool L3 5SR

Tel: 051 794 3270

---

### Project Researchers

Pamela Wadsworth (from 1989)

Derek Bell (from 1989)
Ken Longden (from 1989)
Adrian Hughes (1989)
Linda McGuigan (from 1989)
Dorothy Watt (1986-89)

---

### Associated Researchers

John Meadows
(South Bank Polytechnic)

Bert Sorsby
John Entwistle
(Edge Hill College)

---

### LEA Advisory Teachers

Maureen Smith (1986-89)
(ILEA)

Joan Boden
Karen Hartley
Kevin Cooney (1986-88)
(Knowsley)

Joyce Knaggs (1986-88)
Heather Scott (from 1989)
Ruth Morton (from 1989)
(Lancashire)

PRIMARY SPACE PROJECT
RESEARCH REPORT

January 1990

# Light

by
JONATHAN OSBORNE, PAUL BLACK,
MAUREEN SMITH and JOHN MEADOWS

LIVERPOOL UNIVERSITY PRESS

First published 1990 by
Liverpool University Press
PO Box 147, Liverpool L69 3BX

Reprinted, with corrections, 1992



British Library Cataloguing in Publication Data
Data are available
ISBN 0 85323 466 3

Printed and bound by
Antony Rowe, Limited, Chippenham, England

***CONTENTS***

## *1. INTRODUCTION*

*This introduction is common to all SPACE topic reports and provides an overview of the Project and its programme.*

*The Primary SPACE Project is a classroom-based research project which aims to establish*

- *the ideas which primary school children have in particular science concept areas*
- *the possibility of children modifying their ideas as the result of relevant experiences.*

*The research is funded by the Nuffield Foundation and is being conducted at two centres, the Centre for Research in Primary Science and Technology, Department of Education, University of Liverpool and the Centre for Educational Studies, King's College, London. The joint directors are Professor Wynne Harlen and Professor Paul Black. The Project has one full-time researcher, based in Liverpool, and is supported by a range of other personnel (refer to Project team). Three local education authorities are involved: Inner London Education Authority, Knowsley and Lancashire.*

*The Project is based on the view that children develop their ideas through the experiences they have. With this in mind, the Project has two main aims: firstly, to establish (through an elicitation phase) what specific ideas children have developed and what experiences might have led children to hold these views; and secondly, to see whether, within a normal classroom environment, it is possible to encourage a change in the ideas in a direction which will help children develop a more 'scientific' understanding of the topic (the intervention phase).*

*Eight concept areas have been studied:*

*Electricity*
*Evaporation and condensation*
*Everyday changes in non-living materials*
*Forces and their effect on movement*
*Growth*
*Light*
*Living things' sensitivity to their environment*
*Sound.*

*The Project has been run collaboratively between the University research teams, local education authorities and schools, with the participating teachers playing an active role in the development of the Project work.*

*Over the three year life-span of the Project a close relationship has been established between the University researchers and the teachers, resulting in the development of techniques which advance both classroom practice and research. These methods provide opportunities, within the classroom, for children to express their ideas and to develop their thinking with the guidance of the teacher, and also help researchers towards a better understanding of children's thinking.*

***The involvement of teachers***

*Schools and teachers were not selected for the Project on the basis of a particular background or expertise in primary science. In the majority of cases, two teachers per school were involved, which was advantageous in providing mutual support. Where possible, the Authority provided supply cover for the teachers so that they could attend Project sessions for preparation, training and discussion, during the school day. Sessions were also held in the teachers' own time, after school.*

*The Project team aimed to have as much contact as possible with the teachers throughout the work to facilitate the provision of both training and support. The diversity of experience and differences in teaching style which the teachers brought with them to the Project meant that achieving a uniform style of presentation in all classrooms would not have been possible, or even desirable. Teachers were encouraged to incorporate the Project work into their existing classroom organisation so that both they and the children were as much at ease with the work as with any other classroom experience.*

***The involvement of children***

*The Project involved a cross-section of classes of children throughout the primary age range. A large component of the Project work was classroom-based, and all of the children in the participating classes were involved as far as possible. Small groups of children and individuals were selected for additional activities or interviews to facilitate more detailed discussion of their thinking.*

***The structure of the Project***

*For each of the eight concept areas studied, a list of concepts was compiled to be used by researchers as the basis for the development of work in that area. These lists were drawn up from the standpoint of accepted scientific understanding and contained concepts which were considered to be a necessary part of a scientific understanding of each topic. The lists were not necessarily considered to be statements of the understanding which would be desirable in a child at age eleven, at the end of the Primary phase of schooling. The concept lists defined and outlined the area of interest for each of the studies; what ideas children were able to develop was a matter for empirical investigation.*

*Most of the Project work can be regarded as being organised into four phases, preceded by an extensive pilot phase. These phases are described in the following paragraphs and were as follows:*

*Pilot work*
*Phase 1: Exploration*
*Phase 2: Pre-Intervention Elicitation*
*Phase 3: Intervention*
*Phase 4: Post-Intervention Elicitation*

***The phases of the research***

*Each phase, particularly the Pilot work, was regarded as developmental; techniques and procedures were modified in the light of experience. The modifications involved a refinement of both the exposure materials and the techniques used to elicit ideas. This flexibility allowed the Project team to respond to unexpected situations and to incorporate useful developments into the programme.*

*There were three main aims of the Pilot phase. Firstly, to trial the techniques used to establish children's ideas; secondly, to establish the range of ideas held by primary school children; and thirdly, to familiarise the teachers with the classroom techniques being employed by the Project. This third aim was very important since teachers were being asked to operate in a manner which, to many of them, was very different from their usual style. By allowing teachers a 'practice run', their initial apprehensions were reduced, and the Project rationale became more familiar. In other words, teachers were being given the opportunity to incorporate Project techniques into their teaching, rather than having them imposed upon them.*

*In the Exploration phase children engaged with activities set up in the classroom for them to use, without any direct teaching. The activities were designed to ensure that a range of fairly common experiences (with which children might well be familiar from their everyday lives) was uniformly accessible to all children to provide a focus for their thoughts. In this way, the classroom activities were to help children articulate existing ideas rather than to provide them with novel experiences which would need to be interpreted.*

*Each of the topics studied raised some unique issues of technique and these distinctions led to the Exploration phase receiving differential emphasis. Topics in which the central concepts involved long-term, gradual changes, e.g. 'Growth', necessitated the incorporation of a lengthy exposure period in the study. A much shorter period of exposure, directly prior to elicitation was used with 'Light' and 'Electricity', two topics involving 'instant' changes.*

*During the Exploration, teachers were encouraged to collect their children's ideas using informal classroom techniques. These techniques were:*

i. *Using log-books (free writing/drawing)*

*Where the concept area involved long-term changes, it was suggested that children should make regular observations of the materials, with the frequency of these depending on the rate of change. The log-books could be pictorial or written, depending on the age of the children involved, and any entries could be supplemented by teacher comment if the children's thoughts needed explaining more fully. The main purposes of these log-books were to focus attention on the activities and to provide an informal record of the children's observations and ideas.*

ii. *Structured writing/drawing*

*Writing or drawings produced in response to a particular question were extremely informative. This was particularly so when the teacher asked children to clarify their diagrams and themselves added explanatory notes and comments where necessary, after seeking clarification from children.*

*Teachers were encouraged to note down any comments which emerged during dialogue, rather than ask children to write them down themselves. It was felt that this technique would remove a pressure from children which might otherwise have inhibited the expression of their thoughts.*

iii. *Completing a picture*

*Children were asked to add the relevant points to a picture. This technique ensured that children answered the question posed by the Project team and reduced the possible effects of competence in drawing skills on ease of expression of ideas.*

iv. *Individual discussion*

*The structured drawing provided valuable opportunities for teachers to talk to individual children and to build up a picture of each child's understanding.*

*It was suggested that teachers use an open-ended questioning style with their children. The value of listening to what children said, and of respecting their responses, was emphasised as was the importance of clarifying the meaning of words children used. This style of questioning caused some teachers to be concerned that, by accepting any response whether right or wrong, they might implicitly be reinforcing incorrect ideas. The notion of ideas being acceptable and yet provisional until tested was at the heart of the Project. Where this philosophy was a novelty, some conflict was understandable.*

*In the Elicitation phase, the Project team collected structured data through individual interviews and work with small groups. The individual interviews were held with a random, stratified sample of children to establish the frequencies of ideas held. The same sample of children was interviewed pre- and post-Intervention so that any shifts in ideas could be identified.*

*The Elicitation phase produced a wealth of different ideas from children, and led to some tentative insights into experiences which could have led to the genesis of some of these ideas. During the Intervention teachers used this information as a starting point for classroom activities, or interventions, which were intended to lead to children extending their ideas. In schools where a significant level of teacher involvement was possible, teachers were provided with a general framework to guide their structuring of classroom activities appropriate to their class. Where opportunities for exposing teachers to Project techniques were more limited, teachers were given a package of activities which had been developed by the Project team.*

*Both the framework and the intervention activities were developed as a result of preliminary analysis of the Pre-Intervention Elicitation data. The Intervention strategies were:*

*(a) Encouraging children to test their ideas*

*It was felt that, if pupils were provided with the opportunity to test their ideas in a scientific way, they might find some of their ideas to be unsatisfying. This might encourage the children to develop their thinking in a way compatible with greater scientific competence.*

*(b) Encouraging children to develop more specific definitions for particular key words*

*Teachers asked children to make collections of objects which exemplified particular words, thus enabling children to define words in a relevant context, through using them.*

*(c) Encouraging children to generalise from one specific context to others through discussion.*

*Many ideas which children held appeared to be context-specific. Teachers provided children with opportunities to share ideas and experiences so that they might be enabled to broaden the range of contexts in which their ideas applied.*

*(d) Finding ways to make imperceptible changes perceptible*

*Long-term, gradual changes in objects which could not readily be perceived were problematic for many children. Teachers endeavoured to find appropriate ways of making these changes perceptible. For example, the fact that a liquid could 'disappear' visually and yet still be sensed by the sense of smell - as in the case of perfume - might make the concept of evaporation more accessible to children.*

*(e) Testing the 'right' idea alongside the children's own ideas*

*Children were given activities which involved solving a problem. To complete the activity, a scientific idea had to be applied correctly, thus challenging the child's notion. This confrontation might help children to develop a more scientific idea.*

*In the Post-Intervention Elicitation phase the Project team collected a complementary set of data to that from the Pre-Intervention Elicitation by re-interviewing the same sample of children. The data were analysed to identify changes in ideas across the sample as a whole and also in individual children.*

*These four phases of Project work form a coherent package which provides opportunities for children to explore and develop their scientific understanding as a part of classroom activity, and enables researchers to come nearer to establishing what conceptual development it is possible to encourage within the classroom and the most effective strategies for its encouragement.*

***The implications of the research***

*The SPACE Project has developed a programme which has raised many issues in addition to those of identifying and changing children's ideas in a classroom context. The question of teacher and pupil involvement in such work has become an important part of the Project, and the acknowledgement of the complex interactions inherent in the classroom has led to findings which report changes in teacher and pupil attitudes as well as in ideas. Consequently, the central core of activity, with its pre- and post-test design, should be viewed as just one of the several kinds of change upon which the efficacy of the Project must be judged.*

*The following pages provide a detailed account of the development of the Light topic, the Project findings and the implications which they raise for science education.*

## *2. PREVIOUS RESEARCH*

The child's understanding of phenomena associated with light has attracted considerable interest from a number of researchers in different countries. Work has been done in France, Great Britain, New Zealand, Sweden, Switzerland, India and the USA. Predominantly this work has been done with secondary age pupils. A remarkable feature of this work is the similarity of findings reflecting a cultural and linguistic independency.

Some of the earliest research findings reported in this area are by Piaget (1974) who noted that young children make no connection between eye and object whilst at a later stage they commonly think of vision as ' a passage from the eye to the object'. The nature of this link and association was studied further by Guesne (1976) who has continued to research this area in further depth (Guesne, 1981, 1985). The following is a summary of hers and others findings.

### *2.1 Children's ideas about light*

A minority of children have been found to hold the view that light is an omnipresent medium and which they do not identify with a particular source (La Rosa et .al, 1985). Children holding this conception will view daylight as providing a 'sea of light' which enables vision (Guesne 1978). The light is located in space between source and effect. Such children do not readily acknowledge that light is the result of a disturbance which propagates through space with rectilinear properties.

Guesne(1978) found that a minority older children (13-14) would recognise the notion of light moving in a rectilinear path and use this to explain shadows. She distinguished two distinctive notions which she saw as part of a developmental process. Those children, generally younger (11-12), who equated light with its source, its effects or its state, and and those who recognised light as a separate entity, situated in space between the source and effects that produce it. The former would often talk about light being 'in the bulbs' or 'on the ceiling' whilst the latter would talk about light not being 'able to *pass through* the paper' causing a shadow. However their conception of movement along the path was not clear and their answers suggested that light needs an impetus to maintain its motion throughout space. Stead and Osborne (1980) using a set of simple but revealing multiple choice questions showed that the impetus notion was widely held. With faint sources, the light did not move beyond the surface of the source. It would also travel further at night. This view is supported by work done by Guesne (1978) and Andersson and Karrqvist (1983). There is little evidence in any of the work to support the view that many children commonly see light as something which propagates indefinitely through space.

Shadows are often seen as being 'reflections' of objects (Guesne, 1978). She gives several examples in which the term 'reflection' is used to explain shadow formation. However both she and Ramadas (1989) point to the fact that the term is used loosely to describe the similarity of form i.e that the child is merely equating the light with its effects, noting the correlation of the effects. Such children would be able to correctly predict the shape of the shadow but would not be able to predict the effect of changing the spatial relationship between source, object and screen.

Image formation by a plane mirror has also been extensively investigated (Jung 1981; Guesne, 1985; Watts, 1985; Goldberg and McDermott, 1986). The common feature of this work is the widely and strongly held view that the image is resident on the screen or possibly just behind it. Ramadas (1989) reports a study carried out in India with a group of 14-15 year olds. A teaching sequence which was designed to challenge their ideas about the position of the image notably failed to produce any significant shift in their thinking.

### *2.2 Children's Ideas about vision*

Vision is perceived as an active process in which the subject is the origin of the process. Typically a child will state

> '..Here *my eyes can go right up to the box*......It's my sight.....If it [the box] was fifteen kilometres away, I couldn't see it, because... my sight isn't strong enough...'
>
> Guesne (1978), p 188

Guesne draws a careful distinction between this and historical parallels, notably the Pythagorean view that vision was exclusively due to an invisible fire emanating from the eyes (Ronchi, 1970). Children see the movement from the eyes to the object as something which is essentially abstract and this is clearly differentiated from the 'visual fire' of early theories. She found that for a significant number of children, vision is represented as a process in which the eye is sending out 'rays' which return to the head with a message or picture. Remarkably similar ideas are found in the research of Andersson and Karrqvist (1983) who looked at the understanding of light held by Swedish pupils, age 12-15. Both Guesne and Andersson and Karrqvist (1983) found that the physicist's model is relatively rare at this age which would support Piaget's view that only children who achieve formal operations are capable of recognising that light exists as an independent entity.

Work by Crookes and Goldby (1984) revealed that some children held the view that light comes to the eye and then goes to the object. This idea is also supported by Ramadas & Driver (1989). The latter's research uses some data from the Assessment of Performance Unit in the UK to produce an extensive schema for classifying children's ideas of vision. A simpler version is used by Guesne which identified four common conceptions which she saw as stages in the progression of children's thinking about vision: the notion of ambient light which fills the space providing the light necessary to see; a source-object link which illuminates the object viewed; a source-object link with 'active' vision and a source-object link with receptive vision. Clearly identifiable here are the two separate links that have to be made i.e source-object and object-eye and the joint association between the two. Children who express the ideas represented by the third instance could be trying to reconcile their view that light is needed for vision with the idea that sight is 'active'. A possible obstacle for children developing a scientific understanding, particularly for non-luminous objects, is the recognition that objects reflect light.

Both Guesne (1985) and Andersson and Karrqvist (1981) have pointed to the influence of the metaphors used in language that help to reinforce the idea of vision as a process in which something emanates from the eye. We 'look daggers', 'cast our gaze', have 'piercing eyes' and 'stare at objects'. Such language clearly reflects and reinforces the intuitive understanding found in many children. In addition, comic strip figures have 'X ray vision' which can penetrate walls. Jung (1987) makes the case that common discourse is a more accurate interpretation of children's understandings. Their ideas and language are rooted in a phenomenological approach to learning. La Rosa et .al (1985) used Jung's earlier work (1982) to define an interpretative framework for analysing data from 63 secondary school students (16-17). Their work confirmed the finding of other researchers leading them to the conclusion that 'it is difficult to find situations which challenge the predictive power' of such models.

***References***


Andersson, B. & Karrqvist, C. 1983 How Swedish Pupils, aged 12-15 years, understand light and its properties. *European Journal of Science Education,*5,4,387-402 .

Andersson, B. & Karrqvist, C. 1981 *Light and its Properties.* EKNA report No 8, Institutionen for Praktisk Pedagogik, Goteborg, Universitet, Sweden.

Crookes, J. & Goldby, G. 1984 *How we see things: An introduction to light.* Science Process Curriculum Group, The Science Curriculum review, Leicestershire.

Goldberg, F. 1986 Student Difficulties in Understanding Image Formation by a Plane Mirror. *The Physics Teacher* 24, 472-480.

Guesne, E. 1976 Lumiere et vision des objets: une exemple de representations des phenomenes physique, pre-existant a l'enseignement. Proceedings of GIREP. Taylor and Francis

Guesne, E. 1984 New Trends in Physics Teaching, IV. UNESCO, Paris, France, 179-192.

Guesne, E. 1985 in Driver R., Guesne E. & Tiberghien A. *Childrens' Ideas in Science.* Open University Press. Milton Keynes.

| | | |
|---|---|---|
| Jung, W. | 1981 | Conceptual frameworks in elementary optics. Paper presented at the International Workshop, Selbstverlag Pädagogische Hochschule, Ludwigsburg. |
| Jung, W. | 1987 | *Understanding Student's Understandings: The Case of Elementary Optics*. Proceedings of the Second International Seminar: Misconceptions and Educational Strategies in Science and Mathematics, Vol. III, pp. 268-277. |
| La Rosa, C. Mayer, M. Patrizi, P. & Vincentini-Missoni, M. | 1984 | Commonsense Knowledge in Optics: Preliminary results of an investigation into the properties of light. *European Journal of Science Education,* 6,4,387 -397. |
| Piaget, J. | 1974 | *Understanding Causality*. W.W Norton, New York. |
| Ramadas, J. & Driver, R. | 1989 | *Aspects of Secondary Student's ideas about light.* University of Leeds. |
| Ronchi, V. | 1970 | *The Nature of Light: An Historical Survey.* Heinemann, London. |
| Stead, B.F. & Osborne, R.J. | 1980 | Exploring science students' concepts of light. *Australian Science Teachers Journal,* 26,3, 84-90. |

## *3. METHODOLOGY*

### *3.1 Sample*

*i) Schools*

Five schools from the London area were chosen for this research, all in the Inner London Education Authority. The schools were all primary schools located south of the river. It was the intention to make full use of the whole age range in this study but the difficulties in obtaining a part-time researcher led to a limitation to the junior school age range, that is 7 to 11 year olds.

The selection of the schools was done by the researcher, Maureen Smith, who had already been working in the locality providing support to primary schools in the development of primary science work. Names of the participating teachers, their schools and head teachers are in Appendix 1.

*ii) Teachers*

The teachers invited to participate in the project were those known to the researcher from her previous work. This was advantageous in providing a pre-existing relationship and link between researcher and teachers which could be developed. Teachers were able to use this relationship to express their uncertainties about the work and ask for clarification. Unfortunately, the local authority was unable to release any of the teachers due to the difficulties experienced during this phase in obtaining any supply cover. This meant that all meetings had to take place during the teachers' own time, after school, and this had the effect of curtailing the extent of the teacher contribution to the research on this topic. The number of teachers involved in each school varied between one and three, though in most schools there were two teachers involved in the work. This was useful in that they were able to provide mutual support.

The teachers' normal style of working varied, between individuals who made sole use of classrooms based around groups teaching through topics and an 'integrated' day, and those who preferred to keep the class working together on a common theme. Teachers were encouraged to integrate the activities into their existing mode of working as there was a limitation to the amount of changes that could be expected of them. Many of the difficulties experienced and expressed by teachers were often to do with a lack of confidence in their own understanding of the topic.

*iii) Children*

Despite the limitation to a particular locale, the schools used reflect the wide variation seen in the London area between schools based in deprived areas and those with a substantial middle-class catchment area. Hence the children used in the sample represent children with a wide range of ability and ethnic background. All children in the classes of the participating teachers who were involved in the project to some extent were used for the pre- and post-intervention elicitation activities.

*iv) Liaison*

The part-time research co-ordinator also worked as a science advisory teacher and was able to use this role to provide enhanced support and guidance to the teachers involved in the project. In addition, it provided her with ready access to the schools to trial activities and materials being developed with the research team.

### *3.2 The Research Programme*

Classroom work on the topic of 'light' took place over a relatively long period in the school year which can be summarised as follows.

| | |
|---|---|
| Pilot Exploration | April-June 87 |
| Pre-Intervention Data Collection | Sept-October 87 |
| Intervention | Jan-Feb 88 |
| Post-Intervention Data Collection | March 88 |

The pilot exploration phase was based on interviews with a small number of children. These used a wide range of questions to explore the nature of children's understandings of the topic of light and associated concepts. In addition, drawings and answers to written questions were employed to examine how valuable and reliable such sources were for eliciting children's meanings and understanding. Sample questions are shown in Appendix 2. The exploratory nature of this phase was necessitated by the lack of any substantial literatureappropriate to this age range providing a reference point for the level and depth of children's understanding. Many of the tools devised for probing children's ideas were modifications of methods that had been used with older children. At the end of this phase, the data was examined to determine which were the most valuable lines of approach for eliciting children's ideas about this topic. The other valuable feature of this phase, was that it provided time for developing a relationship with the teacher and the children so that they could become accustomed to the mode of working required.

Essentially, the classroom elicitation techniques were refined by the previous process and it provided an opportunity for teachers and researchers to develop familiarity with the material and each other. Data on children's ideas was then collected from children in classrooms using the selected activities. These questions and activities are shown in Appendix 3. The main methods of elicitation relied on written answers and children's drawings. These were also supplemented by interviews with a few children to provide further insight. No attempt was made to collect interview data systematically from a large number of children. This limitation was imposed by the exigencies of funding and restricted time available from the part-time researcher.

The intervention activities were designed in consultation with the teachers and from an examination of the data collected previously. The data suggested three areas of interest for possible conceptual development and a framework of activities was designed which could be used by children to test their ideas on the behaviour of light. This was not presented as a prescriptive framework, but simply a range of exercises which could be used by children. Teachers and children were free to try other lines of investigation they wished to pursue. After the completion of the intervention phase, another set of elicitations was used with the children based on a similar questions to those used in the elicitation prior to the intervention.

### *3.3 Defining 'Light'*

Any attempt to develop a child's concepts needs to have a map of what a preferred understanding would be. The following list was compiled by the team to provide a map of ideas considered an *a priori* necessity for the development of the scientist's world view.

1. Light travels

2. Light normally travels in straight lines and can be represented by lines.

3. Light is produced by a range of sources and travels outward from the sources.

4. Many objects reflect or re-emit light as well as mirrors.

5. Primary sources of light emit light which travels long distances till it interacts with matter.

6. Vision occurs because light enters the eye from the object.

7. Shadows occur because the light is blocked by the object from travelling. A shadow should be seen as a lack of light rather than a 'reflection of' the object.

This list represents a basis or platform for the fuller understanding of the scientist. The child who thinks that vision occurs by rays emanating from the eye cannot understand the formation of an image in a mirror. Many secondary teachers take such notions for granted, assuming that there is a basic simplicity about these ideas which all children must appreciate. Consequently, this list acts as an ideal reference point; a collection of ideas that children *may* develop by age 11. One of the purposes of the research would be to examine to what extent such ideas do develop in children as a result of their experiences and activities.

These ideas also provide a framework for examining children's ideas allowing three questions to be addressed.

a) How disparate are the conceptions held by many children from such a framework?

b) What development is observable in children's ideas across the age range?

c) What potential does such an intervention have for the development of children's ideas towards this view?

This list was also used as a reference point for the development of the intervention. Given such a framework of ideas, the task was to develop activities which would assist the formation of a fuller understanding in children. The activities were devised using simple materials familiar to children. Their primary role was to provide a focus for discussion of children's thinking and to challenge their existing ideas. Other considerations in designing the activities were that the materials should be simple, easy to manipulate and safe to handle.

***Further Research with Infants***

Further research was undertaken with two classes of infant children in the summer of 1989. The study used similar methods to elicit these children's understanding of light and to explore the effects of the intervention strategies and materials. The results and data obtained from this study will be published as a supplementary report as part of this series of publications in due course.

## *4. PRE-INTERVENTION ELICITATION WORK*

### *4.1 The Pilot Phase*

This phase of the work was carried out by the research team. Ideally it would have been preferable to train the teachers involved to do more of this work. However, the lack of possible release provided little opportunity to do undertake such training. It was decided to use the available time for training the teachers for the main intervention work and for the preliminary exploration, to collaborate with teachers in the classroom .

Using the previously established framework, a set of activities was devised which would provide an opportunity to explore children's ideas about light and its behaviour. These were

a. Using the lights in the room

Children's attention was directed towards the lights and they were asked what the lights were for. The common response 'to see with' was used for further questioning to ask 'How the lights helped us to see' and whether children would describe light as rays that are travelling.

b. Shining a small torch at

- the eyes:

Since this has a physical effect on the eyes is to cause the iris to contract, it was felt that this may provide an experience that the cause was external i.e the torch and the effect was the consequence of too much light entering the eye.

- a piece of paper and moving the torch:

The torch was use to produce a pool of light on a sheet of paper. Moving the torch closer caused the pool to become brighter and vice-versa. This experience was used to explore children's reasoning and models about light.

- a mirror in front of the child with the torch held behind.

The children were asked to use the mirror to see the torch behind them. This did not cause them particular difficulty. They were then asked to draw how it was possible to see the light coming from behind them to see what representations of light they would use.

c. Looking with their eyes covered

The purpose of this was to use a simple activity commonly used by children in games as a means of exploring their ideas about perception. Why did covering their eyes make it difficult to see? Did very young children have notions of the existence of external space being dependent on their vision of it?

d. Making shadows with a torch and simple objects

A range of simple objects e.g books, pencils and hands, were used to produce shadows on paper with a small torch. Children were asked what caused the shadows and provided a range of explanations varying from those saw the shadow as a copy of the object to those that involved an understanding of rays and rectilinear propagation.

e. Looking at cameras and photographs

Photographs are something which children see and use from a young age. Children were asked how it was possible to capture the picture forever to see what understanding of light they used in their explanations.

f. Investigating a range of mirrors

Again, mirrors are part of common everyday experience and the light from a torch was reflected into another part of the room and children asked how this was possible. Plastic mirrors were used which cause distortion and children were asked why this happened. Generally, children provided explanations that talked of the light 'bouncing off' the mirror but their explanations of distorted images tended to focus on the bending of the mirror rather than light itself.

g. Examining the effect of spectacles

Many children wear spectacles and such children, or alternatively the researcher, were used as a focus to ask how spectacles assisted people to see. Answers tended to focus on the defects of sight. Typically, children would mention that without glasses everything was fuzzy and glasses made vision clearer rather than show any understanding of the effect of the lens on light.

h. Investigating sources of light e.g torches, candles, the sun

These sources were shown to the children and children asked to talk about the differences and similarities between them. Much of their discussion was about the strength of the sources with an implicit recognition of an overall similarity.

These activities all represented experiences which provided simple concrete experiences which could act as a focus for discussion about light by the children. For instance, the lights in the room would be turned off so that the room became noticeably darker and the children would be asked where the light was coming from now. This provided insight into whether they simply saw the room as bathed in a pool of light which had diminished or whether they were aware that the room was lit by sunlight, even though the sun was not visible at the time.

One important difference between this topic area and others such as 'growth' is that the phenomena are instantaneous. Either the effect is observable or it is not and there is no long term property or effect. For instance, with the phenomenon of growth, an individual can test a belief that 'water is necessary to make it grow' fairly readily. A child can devise a simple experiment with two similar plants where the amount of water supplied to each plant is controlled. Consequently, it is possible to make causal inferences about the phenomenon as a result of perceptible effects. However, it is not so easy to test out the concept that' light stops when it hits a piece of paper.' Light appears to travel instantaneously from one location to another, its path is invisible and difficult to make perceptible. Consequently children often restrict themselves to simple phenomenological descriptions of the observable which makes it much hard to develop the scientist's notion of a 'ray'. Turning on a torch[1] produces light on the paper simultaneously, which is not observed to travel further. In many children's thinking, such events simply 'happen' and did not appear to require an explanation.

### *4.2 Results from Preliminary Explorations*

Data were collected in the preliminary investigation by the researchers using a mixture of interviews and drawings. However, these were not a prescribed set of questions and additional questions were used to probe areas of interest for further clarification. Interviews were normally conducted with small groups of two or three children, mainly for the sake of expediency given the limited amount of time available. However, they also proved valuable in generating discussion between children on the issue when there was disagreement between them. A video recording was also made of one or two of the interviews at this stage for training purposes with teachers.

The early findings at this stage were summarised by simple counts of statements made by pupils. Essentially, pupils' views were:-

a. Light was localised with a particular source which may be the sun, a bulb or a lamp. Light does not travel very far from the source. Secondary sources of light were identified with an object such as a window, glass door or a ceiling. There was little recognition of the primary source of light in such instances.

b. Many saw the moon as a source of light rather than a reflector of the sun's light.

c. Cats can see in the dark because their eyes are bright and they shine. Cat's eyes are special eyes.

---

1 The torches used in these investigations were of the small 'Duracell' type. These have the advantage of providing a bright but small light source which provides sharp rather than diffuse shadows. The mirrors used were of aluminised plastic. Whilst they do not provide such a good image as an ordinary mirror, they are comparatively safe.

d. There was very little notion of light travelling or understanding of how light gets here. Objects were seen to shine and there were mentions of light coming through the wires by electricity.

e. Sight was explained in terms of seeing with the eyes or some observable mechanism such as the pupil. The pupils gets bigger in dim light to help you to see. There was no understanding of the part played by light in vision other than that 'you need light to see with'.

f. Few children were clear what happens to the sun at night.. Explanations were often anthropomorphic or mechanistic. The sun 'gets tired' or is 'turned off' at night.

g. Mirrors 'bounce' or 'reflect' light but there was little understanding of how bicycle reflectors work. They were described as 'glowing'.

h. Pupil drawings of torches and experiments with light showed the objects but representations of light were often restricted to blobs. Very few drawings showed the light entering the eye.

i. Glasses help you to see better or, if there is nothing wrong with your vision, they make everything blurry. There was no understanding shown of the effect of the glass on the light.

Examples of this are shown in the following extracts from interviews.

Interviewer. How does light get here?

*Sam.* *By the light bulb and from the wires.*

Interviewer. From the wires?

*Sam.* *Yes it comes up through the electricity in the wires.*

Interviewer. What happens to the sun at night?

*Sam.* *It changes into the moon.*

*Anne.* *The sun goes out*

The results from this preliminary work were used as a basis for refining and clarifying the elicitation questions and activities. Those activities which clearly failed to produce from children anything about their picture of light were discarded and a limited subset of questions and activities produced. For instance, questions about seeing in the dark and about glasses had failed to produce anything other than purely operational answers of the kind 'you can't see in the dark' or 'glasses help you to see'. Such answers were not revealing and these activities were omitted. The richest source of data was found to be children's drawings which provided a wealth of detail about the models they were using to explain the observed phenomena.

After a preliminary analysis of these findings, a new set of questions and activities was then used by the teachers and the researchers for the first phase of data collection in Sept/October 87. Data from the elicitation phase were then used as a basis for the designing and planning the framework of the intervention.

### *4.3 Elicitation*

This phase of the work was carried out in collaboration with the teachers in the schools from September to October 87. Children seemed to enjoy the opportunity to partake in this work and they produced a large body of data for examination and analysis. Whilst it is inevitable that some cross-fertilisation of ideas would occur, the occasions on which this happened were relatively minimal as children valued the opportunity to express their own ideas in a non-judgemental situation where they were encouraged to express their thinking.

The activities for the initial elicitation were revised in the light of the experiences gained from the pilot phase by the elimination of those activities which had not proved fruitful in providing data about children's understanding of light. Since the pilot phase had shown that children's drawings were a particularly valuable insight into their thinking, the activities designed for the elicitation made extensive use of children's drawing.

Six activities were designed for use with children to explore phenomena associated with light. These were

a. Investigating where light comes from
b. How do bicycle reflectors work?
c. Investigations with a torch and a mirror.
d. Investigations with a torch and paper.
e. Looking at candles.
f. How do we see?

The activities were complemented with a list of activities and questions to be used by teachers and researchers, full details of which are shown in Appendix 3. The list was not prescriptive and children were allowed to explore and follow up their own line of thinking. During this phase, data were collected by the use of:

*Drawings*

Children were asked to use drawings to show what was happening to the light in their investigation and were encouraged to use such drawings as a recording tool. With some activities, to simplify the task, drawings were provided and children asked to add to the drawing to show what was happening e.g a girl looking at a candle.

*Written Material*

Children were encouraged to provide written responses to some questions which provided data that were used by the researchers. Typical questions that children wrote about without difficulty were:

> How does light get here from the sun?
> What happens to light at night?

However, not all children, particularly younger children were happy about writing and generally this was not such a useful method of obtaining an insight into the thinking of children.

*Organisation*

Much of the work in the elicitation phase was done by the researchers working with the teachers. This provided a valuable opportunity for developing the methodology of the project with the teachers. It also provided valuable insights of the children's thinking about the topic for the teachers which a useful preparation for the intervention phase. Some teachers expressed concern about the demands posed in terms of time and how to integrate the work into their existing range of activities. However, this phase of the work allowed time for these issues to be resolved and for the teachers to prepare and contribute to the intervention.

## 5. CHILDREN'S IDEAS

### 5.1 An informal look at childrens' ideas

Examination of the data showed four areas of particular interest in the work produced by children. These were

a. Children's ideas about sources of light

b. Children's representations of light

c. Children's understanding of vision

d. The context dependence of the answers

### 5.2 Sources of Light

Children were asked to draw pictures of all the different things that they think can give off light. The following diagrams show typical drawings provided by children of differing ages.

Fig 5.1 Age 8

Fig 5.2 Age 11

The most interesting feature was the wide range of sources that children were easily able to think of and draw. This range showed little variation with age (Fig 5.1 & 5.2), though as these drawings show, these are predominantly primary sources. Some drawings did include reflectors and windows but these were much rarer. When children did mention secondary sources such as the ceiling, the window or sky, there was generally no further explanation of the nature of the source. A few children would add to their explanation by saying that 'the light is coming through the window from the sun.' thus demonstrating an awareness of the secondary nature of the source. It is possible that the use of the phrase 'give off' in the question, focuses children's thinking on primary sources.

Very few children showed less than three sources and many drew in excess of 6. The other feature was that there was no marked change with age as to the number of sources or the nature of the drawings. This data suggests that the idea of a source of light and an awareness of a wide variety of sources, is a well established concept by the age of 7/8, the youngest age which this study dealt with. Perhaps this is not surprising, because, if children's understandings are based in their perceptions, then everyday life provides a wealth of observations of a range of light sources.

Another notable aspect was the tendency of children to include simple lines on many of the sources, particularly the sun and light bulbs. The reasons for this representation were not clear. When asked why they had used such lines, some children commented that 'that is how you draw the sun'. However, Fig 5.2 is an example where this representation is extended to other light sources and for most children it shows the first representation of light as a line. A partial explanation may lie in the fact that the eye 'sees' lines coming off many point sources, particularly on wet nights or from bulbs with clear glass envelopes due to refraction of the light. In a later investigation with bulbs in the electricity topic, one child remarked 'Look, you can see the light coming off in lines.' This would suggest that the children are simply representing in their drawings an accurate representation of what they observe.

The source most commonly mentioned by children was the sun. The strength of this idea is supported by the exploration of the 'growth' topic where only two children mention a light source other than the sun as being necessary for growth. This is also confirmed by an exercise in which children were asked to write three sentences that included the word 'light'. A large number contained statements of the form 'Light comes from the sun.'

However, children had much more difficulty postulating a mechanism as to how light travels and arrives on earth to explain how light gets here. The most sophisticated would provide explanations of the form

*'The sun beams light down onto the earth'*

or show drawings of the form shown in Fig 5.3.

Fig 5.3 Age 7

Only a few showed clear evidence of a model of light which is travelling (Fig 5.4). Similarly in the research on sound, it was found that young children had no notion of sound travelling - they hear because they listen hard.

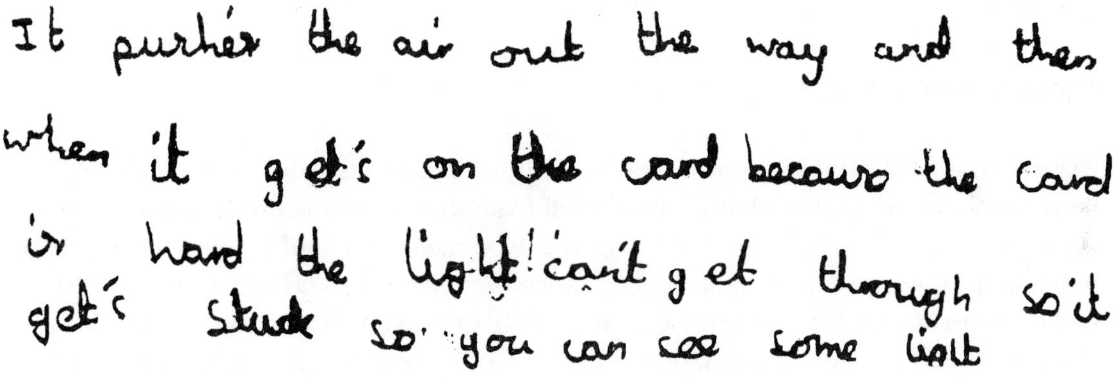

It pushes the air out the way and then when it get's on the card because the card is hard the light can't get through so it get's stuck so you can see some light

Fig 5.4 Age 9

Interestingly, this answer reveals that this child had not yet understood the more difficult concept that not all the light stops at the card.

However, answers of this sophistication were generally rare and it was much more common to provide answers of the form, 'the sun', 'by rays' or 'it beams down'. Many children had no explanation for how the light arrived. No children were found who elaborated on these answers and no evidence was found that the scientist's abstraction of light as a ray, which propagates rectilinearly, is part of children's vocabulary and understanding at this age.

Another interesting feature was the range of children's astronomical ideas that were elicited in response to the question 'What happens to the light at night?'. There seems to be a distinction here between older children, who are developing an understanding that the world turns so that the sun shines on the other side, and younger children who provide other explanations. Typically older children would say

*'The world turns so that the sun is on the other side.'*
*'It comes from the sun and reflects from the moon'*

However, much more common were explanations that stated

*'It goes off.'*

*'The sun goes down and the moon takes its place.'*

*'The clouds cover it.'*

*'It moves to another part of the world because the sun moves.'*

Since these explanations did not reveal much about children's perceptions of light but rather their understanding of the world and the solar system, no attempt was made at any further analysis. However, they are revealing in that they show a range of astronomical models held by children and that there is some development across the age range.

### *5..3 Representations of Light*

Examination of the drawings produced by children showed a wide variation in the representations of light produced by children from that of the mature scientist to none. It is impossible to see a 'light ray' although it can be made perceptible by scattering dust in its path. Hence any attempt at representation was seen as an effort by the child to make the imperceptible, perceptible, that is to concretise an abstraction. In such a situation it was of interest to examine what kind of representation would be used by children to show their thinking about light and its properties.[1]

In the elicitation activity there were five possible questions that required a drawing for an answer where children could have shown some representation of light. These were, showing how they saw the light from a torch placed behind their head in a mirror; how they saw a candle, a book and a clock and showing what happened to light as it travelled through a box with holes in. Clearly this exercise was difficult, particularly for lower junior children where the majority showed no representation. Those that did represent light, most commonly represented it as a line. Fig 5.5 shows a typical example.

---

1 In analysing the data and discussing representations of light, links between the object and the eye were considered to show children's thinking about vision and sight rather than their thinking about light.

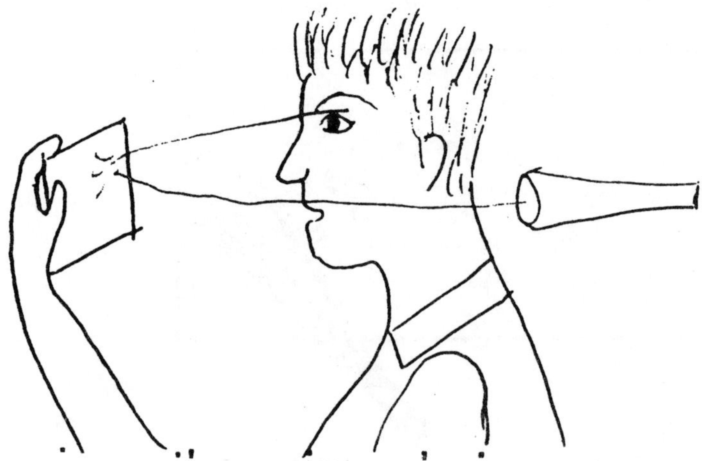

Fig 5.5 Age 9

Another common feature of the children's representations of light was the tendency to show a 'blob' of light on the mirror or torch. The term 'blob' was used by the research team to describe representations of light (on the mirror) such as that in Fig 5.5 and 5.6. Children would also tend to use a combination of 'beam' and 'blob' as in Fig 5.6 which was referred to as a 'dual' representation. The same child in a later question showed light represented as a line which was another example of the context dependence of the answer and dual representations.

Fig 5.6 Age 8

The tendency to use lines may be an extension of the representation of light around sources such as the sun with short lines (Fig 5.1) which has been discussed. Nearly all children used this representation when drawing sources and very few individuals represented light as a line without showing the simple representation as well. If this was so, it could be seen as a pre-cursor or line of development to a more sophisticated model of light.

Other models that were used to represent light were particles. Here the light was drawn as a series of dots, beams of light (Fig 5.7) and occasionally a sea of light which was sketched into the drawing (Fig 5.8).

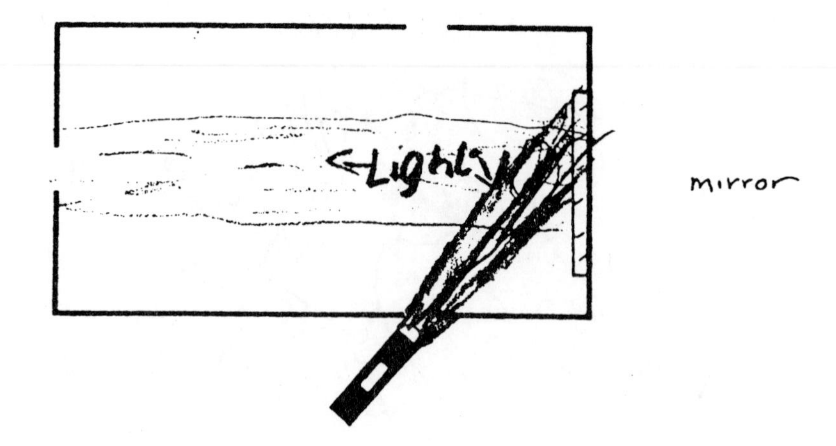


Fig 5.7 Age 11

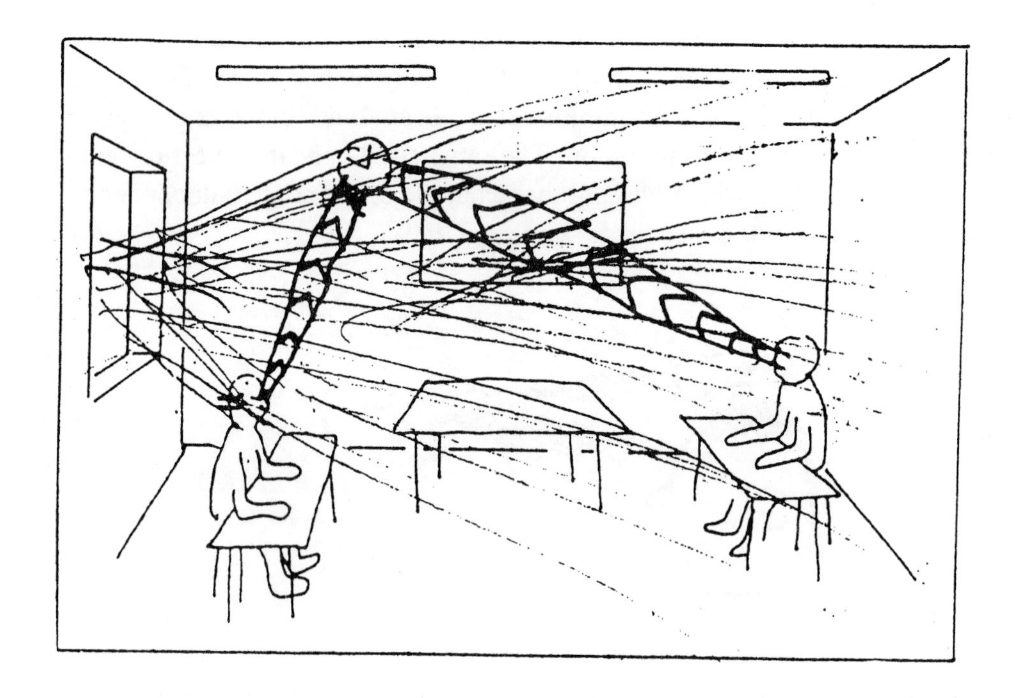

Fig 5.8 Age 11

Fig 5.7 also shows another feature of many children's answer to this question. This was a lack of any understanding of reflection. The drawing is consistent with a model of behaviour for mirrors which bounces light back regardless of the angle of incidence. Only older children were able to show correctly the behaviour of the light when incident on the mirror in this question. Fig 5.8 also indicates an understanding of vision which is discussed in section 7.1c.

Another important feature of the representations shown was the increasing number of children who indicated a specific direction of travel on the diagrams with increasing age (Fig 5.9). No direct request was made for such information and there is no evidence to explain why the direction was included.

However, such representations, typified by Fig 5.9, suggest that children are increasingly seeing light as something which travels and has a direction of travel which would indicate a development in their thinking. Obviously further questioning might reveal such a development and it is unfortunate that there was insufficient time to probe this progression further.

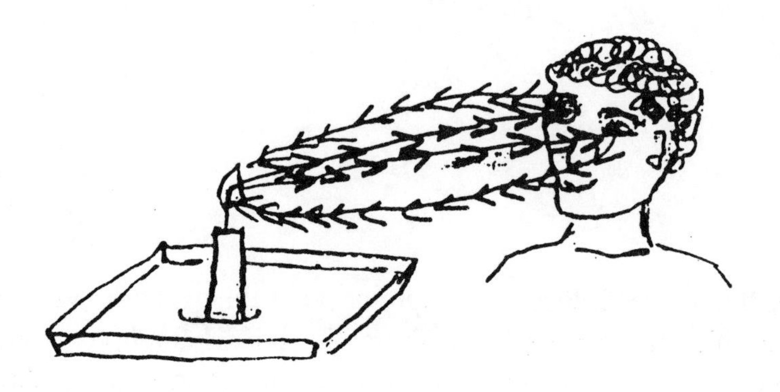

Fig 5.9 Age 11

In summary, a proportion of children were capable of showing a representation of light which was more than a simple representation (Fig 5.2) and the proportion increased with older children. The predominant representation was that of a straight line with several other methods shown, but representations could be context dependent. Older children could more readily indicate a direction of travel which may be an indication of a recognition that light travels.

### *5.4 Children's ideas of vision*

Perhaps one of the most interesting features of this research was the wealth of data it exposed about the range of ideas that children hold about the nature of vision. Essentially, this can be divided into four areas.

- i. No explanation
- ii. Explanations without links
- iii. Explanations with single links
- iv. Explanations with dual links.

#### *i. No explanation*

For many children, particularly younger children, the process of vision appears to be non-problematic in that their drawings and explanations provide no indication of anything other than the simple act of looking. When asked to provide a drawing to show 'How you see a book?', there would be no information, other than the simple observable features (Fig 5.10).

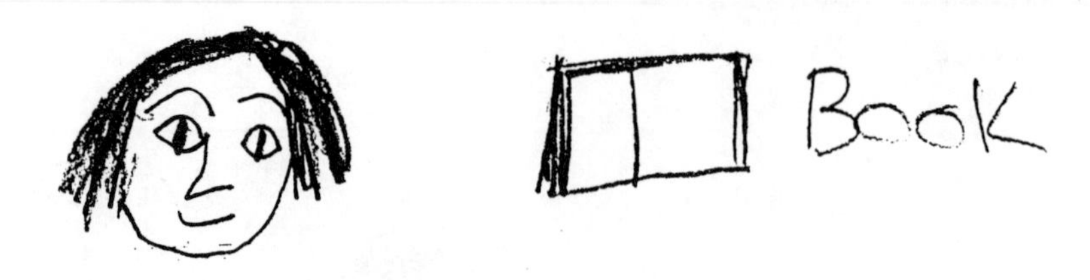


Fig 5.10 Age 11

There was also evidence that some children's interpretation of the question was limited to a descriptive answer as their answers were literal drawings of what a book would look like. However, there were two other questions which were attempting to elicit the nature of the child's understanding of vision which compensated for such interpretations of the question. What was evident was that providing any explanation of viewing objects which are secondary sources of light e.g books, was particularly difficult. Many common-sense explanations and drawings of this form shown in Fig 5.11 were observed.

Fig 5.11 (a) Age 9 Fig 5.11 (b)

*ii. Explanations without links*

For some children, the explanation of how we see an object such as a book, candle or clock was not problematic. The explanation is a simple mechanistic type which recognises that your eyes are essential to vision (Fig 5.12).

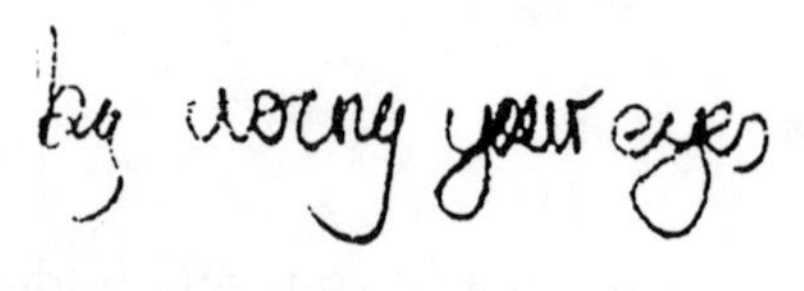


Fig 5.12 Age 11

No further explanation is needed and the impression given by the children was that the rationale is self-evident. Some responses of this form tie the explanation to the pupil of the eye which was seen as being involved in vision (Fig 5.13).

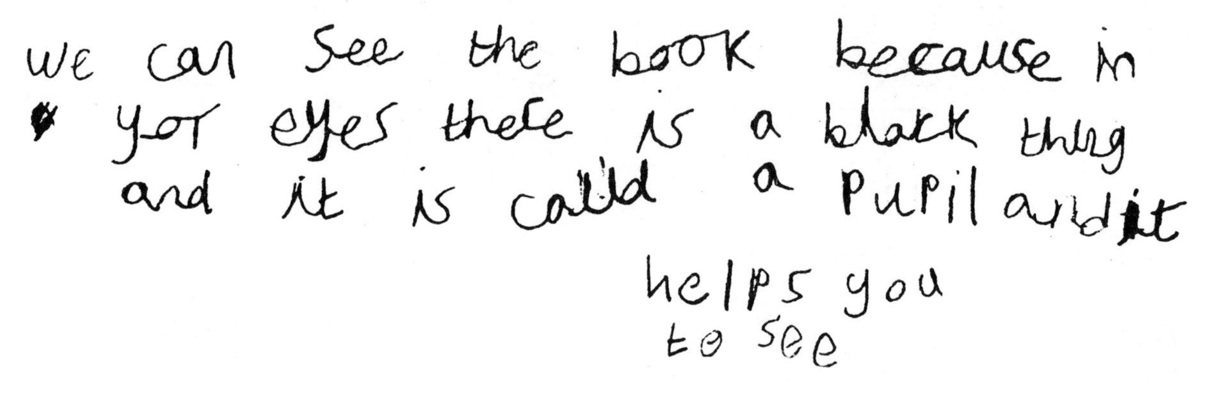


Fig 5.13 Age 8

The other aspect observed in these simple explanations was a recognition that light is needed for vision (Fig 5.14).

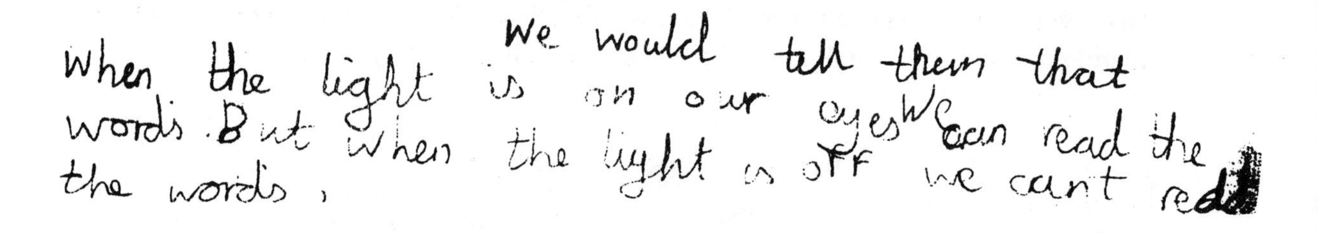


Fig 5.14 Age 10

Such an explanation acknowledges that light is a pre-requisite for vision but fails to provide further detail of the role played by light. In the post intervention elicitation, where children were asked to write three sentences about light, the statement that 'light is needed to see' was commonly expressed.

*iii. Explanations in terms of simple links*

Many children provided explanations or drawings that showed simple links between the eye and the object (Fig 5.15).

(contd overleaf)

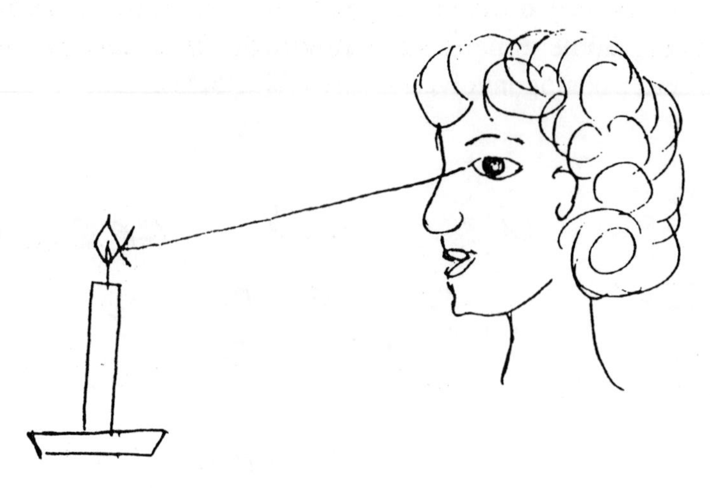

Fig 5.15 Age 10

The large number of such responses show that these children saw vision as an active process. This is not surprising and has been reported elsewhere in the literature. To look at an object, there is an action required of an individual to either move their head or eyes. The vocabulary and the metaphors of the language also imply action so that you 'give looks' i.e 'she gave me a look like daggers' or 'his eyes shone like pearls '. However, though many answers did show a direction, there were also answers which merely recognised the link and did not show any direction (Fig 5.16).

Fig 5.16 Age 8

The majority of such answers showed a representation for vision using lines. However, a few indicated the link in terms of particles (Fig 5.17). There was no evidence that this reflects a view of light but it did show a different conception of the link between eye and object consistent with a particle interpretation.

Fig 5.17 Age 10

Single links directed towards the eye are comparatively rare (Fig 5.18) which reflects the fact that only a very small minority of children had a model of vision which corresponds with the scientist's view.

Fig 5.18 Age 10

*iv. Dual links*

A small but significant number of children recognised the need to show a source-object and object-eye link to explain vision. Identifying that 'light is necessary to see' and that 'we need our eyes to see with', they showed these two factors in a variety of forms. The simplest was that which shows the dual link with no direction (Fig 5.19).

The interesting feature of Fig 5.19 is that it also shows an attempt to reconcile these two ideas with a process of active vision. Light goes (presumably) to the eye and then to the mirror. However, nearly all dual representations showed a direction as well.

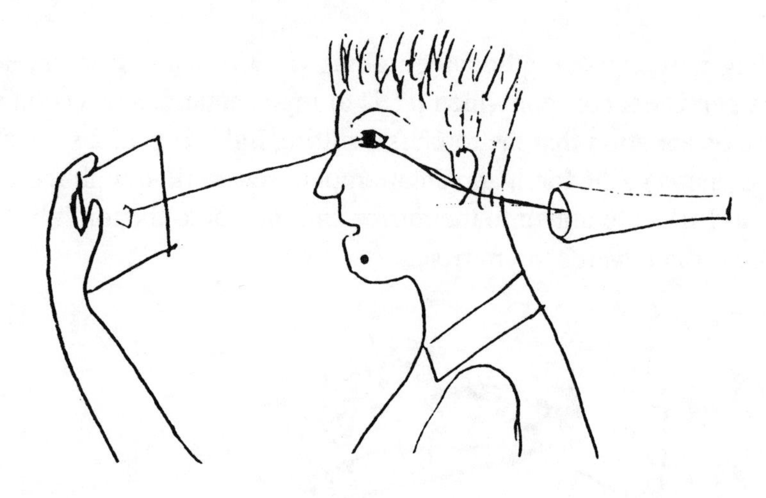

Fig 5.19 Age 11

The most common form of dual representation was one which showed the dual links both directed toward the object (Fig 5.20).

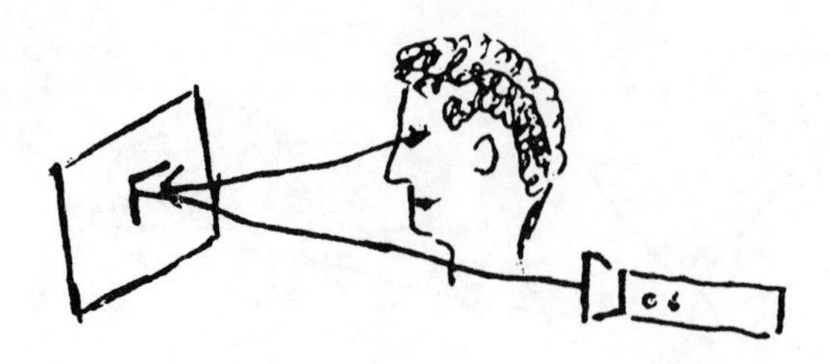

Fig 5.20 Age 10

This representation is a logical expression for children who believe that light 'gets stuck' once it reaches the object and fits with the conception of vision as being 'active'. This term is used to describe children who described and drew vision in terms of lines or rays emanating from the eye. Such a representation is clearly shown in Fig 5.21. Even though the object viewed is a primary source, children's diagrams still showed a representation of 'active vision'.

Fig 5.21 Age 7

However, some diagrams explaining how the light is viewed in a mirror, reveal that 'active vision' is a persistent concept which leads to representations of vision which contradict a simple observation that the torch is emitting light. Fig 5.22 shows an extension of active vision to the torch and an attempt to reconcile it with the emission of light from the torch. Lines are drawn to the mirror and then onto the torch but there is also a line from the torch towards the mirror.

Fig 5.22 Age 9

Another variation which again places pre-eminence on the notion that vision is active whilst recognising that light is necessary for vision, were drawings which showed the light coming to the eye and then passing to the object (Fig 5.23).

Fig 5.23 Age 10

Finally, there were a few children who show a representation consistent with the scientific view (Fig 5.24). Such children were a small minority and there was some evidence that the numbers increased with age.

Fig 5.24 Age 10

Such examples for the book are comparatively rare and there are many more examples of a scientific representation for the torch and mirror (Fig 5.25). The simplest explanation for this would be that observations in the context of the torch and the mirror support the idea that light passes from the source to the mirror to the eye, as it is possible to see the light 'bouncing off' the mirror onto the face. However, there is no evidence to support such an idea with a secondary source of light such as a book.

Fig 5.25 Age 11

### *5.4 Context Dependence.*

The other major feature of the answers and drawings provided by children was the context dependence of the answers. The explanation provided for vision by one child would vary from one question to another within a remarkably short time without any recognition of the of the contradictions that this raises for an adult. There are several possible explanations for this. The simplest would be to say that children saw these situations as instances of different phenomena. A light source shown in Fig 5.26(a) is very different from a book shown in Fig 5.26(b). Consequently the responses provided are different and non-problematic for the child. Similarly Fig 5.27(a) could be considered an instance of 'reflection' whilst Fig 5.27(b) an explanation of 'vision'.

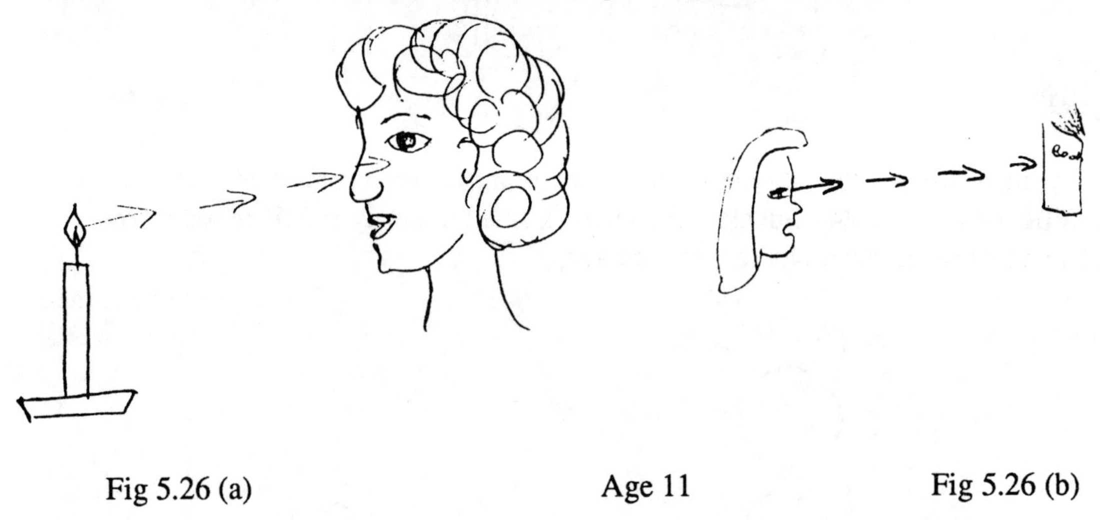

Fig 5.26 (a) Age 11 Fig 5.26 (b)

However, both responses were generated in response to questions about 'How we see'. It is interesting that this child recognised that the candle emitting light is a primary source, which enters the eye whilst with secondary sources, she fell back on the view that vision is active. This would imply that it is impossible for children to develop a scientific view of vision until they are aware that objects such as books are capable of scattering light. Another example of this is shown in Fig 5.27(a) and Fig 5.27(b).

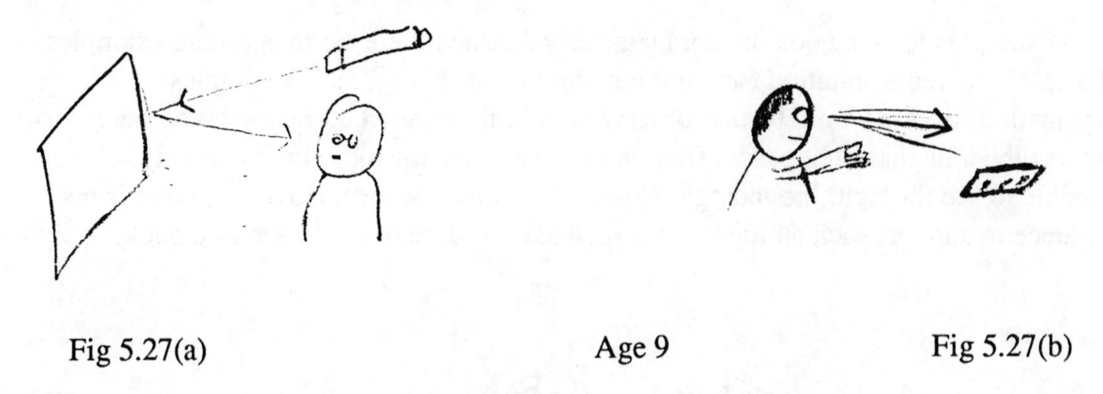

Fig 5.27(a) Age 9 Fig 5.27(b)

Clearly with a luminous source the light comes to the eye but when the source scatters light, the response shows an interpretation that used an explanation of vision as 'active'.

Similar examples can be found for the representations that children use in their drawings. Fig 5.28(a) shows light represented as a line and yet later, on the same occasion, this child used a beam to represent light (Fig 5.28(b)).

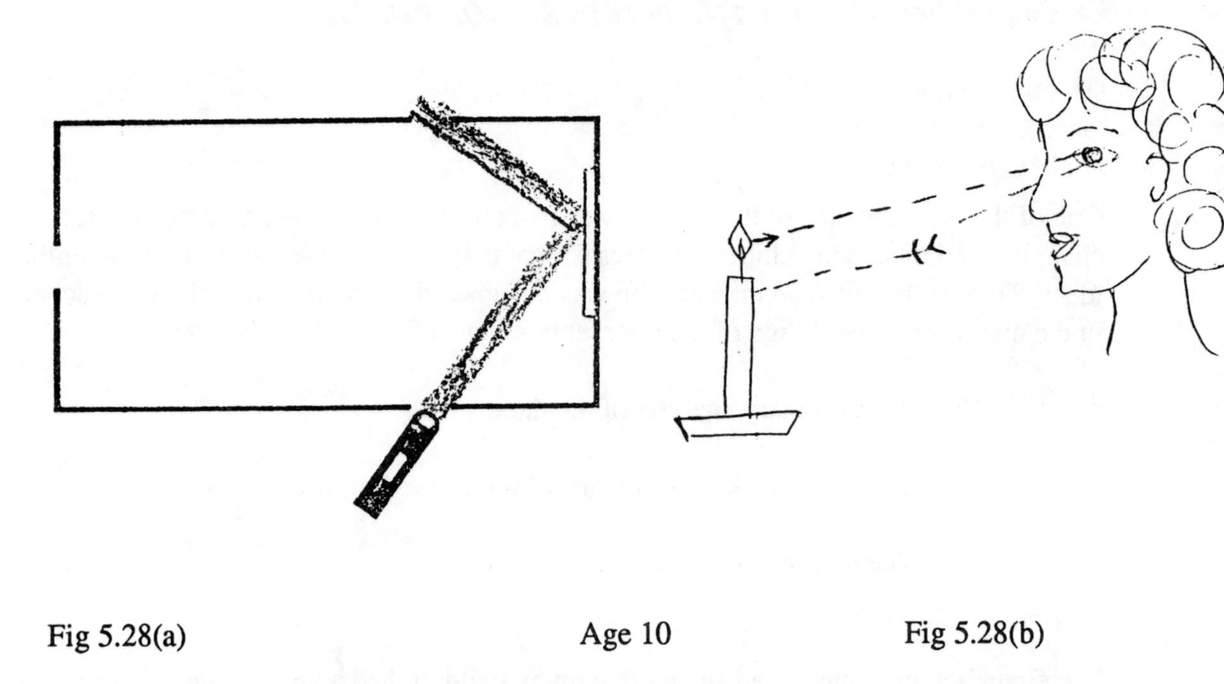

Fig 5.28(a) Age 10 Fig 5.28(b)

The previous examples show that children were using models which are specific to their observations within a particular context. Torches do appear to emit light in beams so the light was drawn as a beam whilst candles do not.

Context dependent responses were observed more often in older children as very rarely had younger children developed more sophisticated models of vision. The implication would be that children's ideas are fluid and lack generalisability, yet are evolving as they get older to incorporate a wider range of observable features which are context dependent. More sophisticated attempts would show object-eye and source-object links which reverted to only one of these links in another response. Table 7.10 shows the number of children using dual models to explain vision which are context dependent and data on this is presented in section 7.1d.

## *6. THE INTERVENTION PHASE*

One of the primary aims of this project was to examine what potential there was for changing and modifying children's ideas so that they can become closer to the scientific ideas. To examine this, an intervention was designed that could be used by the teachers in the classroom. The design of the intervention was influenced by three factors

(a) A preliminary analysis of the data.

(b) The framework of a 'scientific' understanding reviewed earlier.

(c) The teachers' contributions and ideas.

The first elicitation phase had shown that many children had a very limited understanding of light that was rooted in observation. Very few children had a representation of light which consistently approached that of a line or rays and their models of vision were based on explanations that were mechanistic or personal e.g 'You need light to see with', 'You see with your eyes.'

The rationale underpinning the intervention was twofold. Firstly that children should be provided with an opportunity to express their own ideas and hypotheses about phenomena, and that experiences provided by the intervention should be appropriate to developing their thinking from their current conceptions. Secondly, that since one of the purposes of science education is to develop an understanding which is closer to the scientific perspective, intervention activities should facilitate the development of this understanding.

Thus for the design of the interventions, a list of concepts commensurate with a scientific understanding was compiled (3.3). These were then compared with the evidence of the nature of children's thinking observed from the data to guide the design of the interventions so that they would reflect the principles previously outlined. Intervention activities were selected that would require children to hypothesise about the way light travels and how they were able to see light, requiring them to use a representation of light.

It was considered unlikely that any limited intervention would achieve a major shift in children's understanding of vision. Therefore, it was felt that this phase should concentrate on providing experiences which would develop the simplest ideas in the framework; that light travels, and travels in straight lines. Activities which led to the development of these ideas would establish a solid platform for later work. In addition, they might lead incidentally to the growth of a more sophisticated understanding of vision which was more stable and less context-dependent.

Common to all these activities was a requirement that children should discuss and represent their ideas about possible solutions prior to any attempt and for the successful solution to be drawn and discussed with their peers and teacher.

Many of the teachers expressed a desire for simple activities that could be performed by groups within a topic-based approach, focussing on light. This would provide an informal context for teachers to introduce the activities and allow the children to try out their own ideas on a collaborative basis. The teachers were then able to take a non-directive role without losing enthusiasm for the work. Teachers had commented in an earlier meeting that the role expected of them was one that they were not used to. The work required that teachers should be less judgemental than normal and gave children an opportunity to discuss and express their own ideas without criticism from the teacher. In addition, the emphasis placed by the project on conceptual development was one that was unfamiliar to teachers.

Consequently a set of activities was designed for use by teachers that would allow children to test some of their own ideas and develop a wider experience of phenomena associated with light. One of the inherent problems in designing such activities is that the light does not perceptibly travel. As a result, the activities were all designed to help the development of such a hypothesis from their observations. Teachers were asked to avoid 'telling' children that light travelling in straight lines would explain what they see and allow children an opportunity to express and evaluate their own thoughts before contributing such an idea to the discussion.

All of the activities used simple materials and a common process of presenting to the children a simple piece of factual knowledge about light. A problem was then posed to the children who were asked to devise a solution to the problem, sketching their solution first and then testing their idea. Full details can be found in Appendix 4.

*Activity 1: Bouncing Light around a table*

This problem (Fig 6.1) was set in the context of a simple game for the children. Children were reminded that light can be 'bounced off ' i.e reflected from mirrors. The problem was posed as one of 'How could they make the light go round every side of the table?' A strong torch, mirrors and plasticene were provided and the children had to discuss a preliminary solution before attempting this exercise. When they had devised a possible solution,the children would work as a group and use the mirrors, held in position by the plasticene, to test their idea. The mirror angles could be adjusted easily and the light directed from one child to another.

The intention of such an exercise was that it would provide an experience which *may* develop the concept that light travels and goes in straight lines. Children would have to talk about a solution in terms of 'light going from one mirror to another' and implicitly recognising it as a medium which travels. Children's interest in performing this task was generally good though the manipulative skills required were quite demanding.

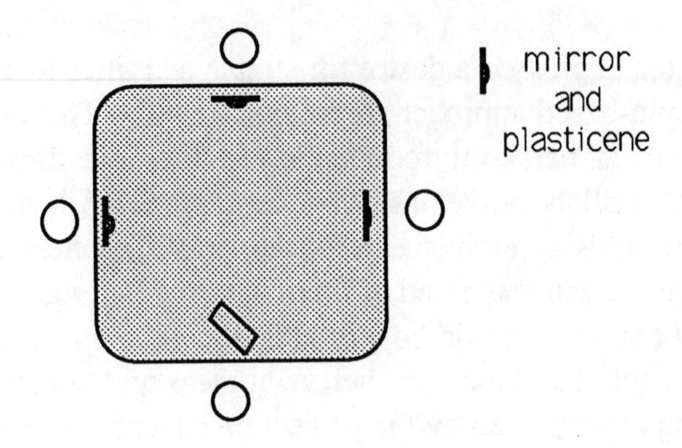


Fig 6.1

*Activity 2: Investigating Shadows.*

Again the intention of this activity was to develop the idea of light travelling through space in straight lines and to encourage the use and development of a method of representing light. The activities were presented as prediction exercises and children were asked to guess and predict the shapes of shadows formed by a variety of objects, to construct a method of testing their ideas and record their results afterwards. Teachers were asked to provide an opportunity for children to use their own ideas, by discussing with the group initially what caused shadows and when did we get lots of shadows. This activity was emphasised as an important process if any conceptual adjustment was to take place. No development of, or conflict with existing models, could occur unless the child was aware of his or her ideas.

*Activity 3: Passing light through boxes.*

This activity made use of shoe boxes with small holes positioned on each side(Fig 6.2). In addition, the box had a mirror placed at one end of it. Children were asked to predict where the light would go when the torch was turned on by adding to Fig 6.2 and then, to repeat this process in a second situation, where the torch was directed at the mirror through the hole in the side. As an activity, they set up the box with the torch in the situations shown and tested whether the light from the torch was visible at the various holes in the box.

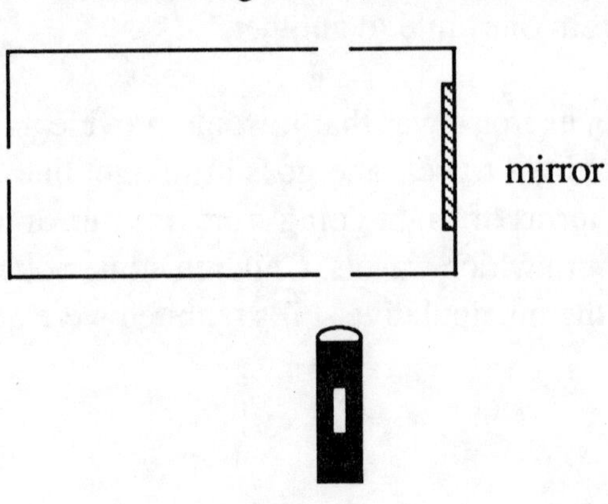


Fig 6.2

When they had completed the exercise, they were asked to draw again where they thought the light went in the box and compare their current thoughts with their previous drawings.

The intervention took place over a month and teachers were asked to try all activities with groups of children when appropriate to their normal classroom work. Teachers received a visit from one of the researchers during this phase of the work to provide support and guidance. In addition, researchers were involved in trying activities with groups of children.

Obviously, there is a methodological problem here in that there is a lack of definitive standardisation. However, one of the available afternoon meetings was spent in introducing the materials to the teachers, providing them with an opportunity to test and evaluate the materials and raise and share problems that might occur in the classroom. In addition, the extended time allowed for such material meant that all teachers had time to complete the interventions with all their pupils. This means that the work is not a tightly controlled research study but possibly provides better data about what is possible in normal situations facing teachers in normal classrooms. Finally, teachers were not restricted to these activities and some extended the work further to investigating light bulbs and work on colour.

The activities are different from a process approach in that they focus on conceptual understanding and its development at a level which would be appropriate to the existing ideas shown by children. In addition, the activities have provided an opportunity for children to test and evaluate their own ideas by experimentation. They represent a limited approach, constrained by the time available, which is a reflection of classroom realities. They do at least have the benefit of being an empirical response to the data gained from the initial elicitation of children and their design was based on an appreciation of the levels of understanding of phenomena associated with light in young children. However, any attempt to judge the value of one activity would be to place too much emphasis on anyone aspect of the intervention which is best judged by the overall and holistic view.

# 7. THE EFFECTS OF THE INTERVENTION

## 7.1 Changes in Children's ideas

This section gives the technical analysis of the data gathered during this study. The main findings are summarised in section 9. Analysis of the data identified three main areas of focus in children's ideas. These were ideas about

a. Sources of Light
b. Representations of Light
c. The nature of Vision
d. Context dependence

This data on children's ideas about light was gathered in two phases, the elicitation phase, prior to the intervention and by a second elicitation exercise after the intervention. This produced a large collection of data for analysis which is presented and discussed in this section. The elicitation activities consisted of activities that were designed to focus and orientate children's thinking on particular phenomena associated with light. Children were then asked to draw and write answers to specific questions about the phenomena and this provided the vast majority of the data. Some data were also collected by interview. Ideally, in the interest of reliability and validity, it would have been preferable to collect much more of the data in the elicitation phase by interview. This would also have provided an opportunity to pursue in further depth some of the interesting responses that children provided to written questions. However, the limits imposed by the staff resources available meant that the research team was constrained to use methods that were broad in their focus, if lacking in depth and rigour. This problem was addressed in two ways. Firstly the study was limited to specific activities that the early exploratory work had shown to generate meaningful and interesting responses from children. Secondly, a substantial amount of redundancy was built into the elicitation activities in order to evaluate the consistency of the responses provided by the children. The issue of consistency is discussed in part (iv) of this section.

One advantage of this strategy is that allowed a reasonably large sample that represented a mix of abilities, school catchment areas and sexes to be used. One constraint that was imposed by circumstances relatively early in the project was a decision to limit the study to junior age children only. Teachers in the schools used were unwilling to involve infant children in the project at this stage till they had developed a body of experience about the methods of the project.

It is also worth noting that substantially more data was collected than shown in the sample sizes given. Data presented here is from elicitation activities with children who were present before and after the intervention. Inevitably, there was some erosion of the sample due to illness and movement of children.

The data sample was sufficiently large to enable analysis of frequency of response by groups and to compare changes in individual responses. The analysis of frequency of response was done using networks . This provided a tool which enabled some clarification and summarisation of the data to take place, together with a quantitative analysis. In addition, it provides a clear, visual representation of the main features of children's thinking at this age.

### *7.1a Sources of Light*

Both elicitation activities had included questions about the origin of light. In particular, the questions asked were

> 'Look around the room. Where do you think the light is coming from?'
> 'Draw pictures of all the different things that you think can give off light.'
> 'How does light get here?'

The wide-ranging responses about sources provided a large body of data about the way light travels as viewed by these children. These were summarised by using network analysis.

The essential method is to examine a large number of the responses and look for what categories of response emerge from the data. Children's ideas about sources were a clear feature of the responses and so it was decided to devote one network to these. In the network shown (Table 7.1), there are two predominant aspects to their responses; ideas about the sources of light and ideas about how light arrives. This is indicated by the use of an inclusive bracket called a 'bra' (Fig 7.1(a)). The use of the 'bra' indicates that any response of a child normally includes both aspects.

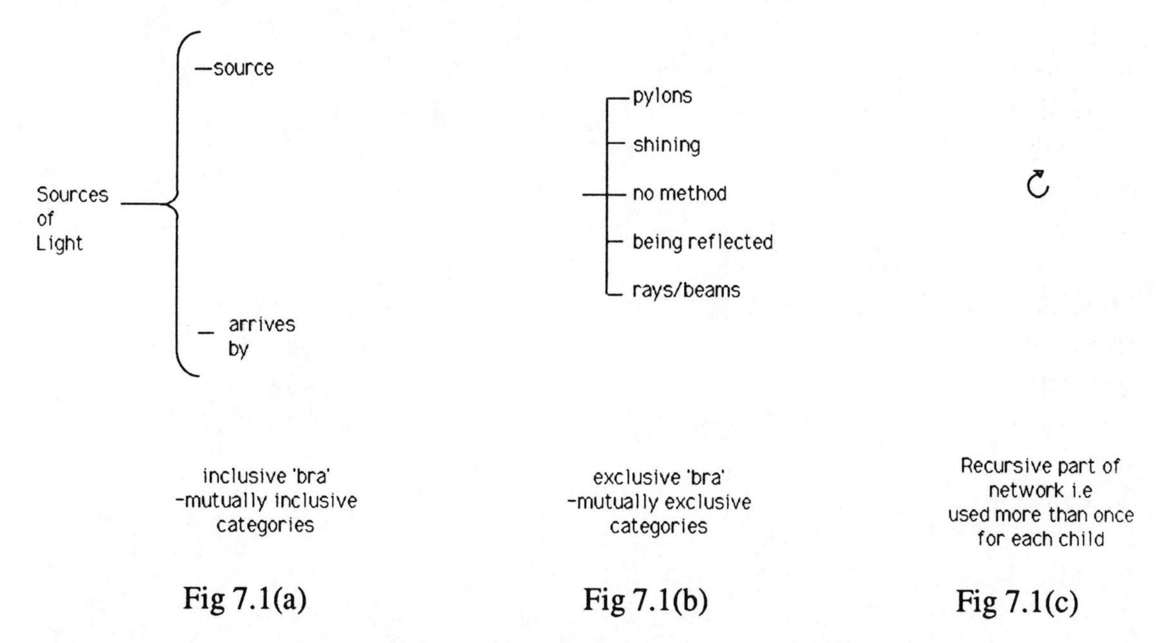


Fig 7.1(a) Fig 7.1(b) Fig 7.1(c)

A further division is then made for children's responses about how the light arrives and their ideas about sources. The division for their idea about the former uses an exclusive bracket called a 'bar' (Fig 7.1(b)). These responses are deemed to be so since children who say that light gets here by pylons do not also say that it gets here by rays or beams. Included within this piece of the network is a separate terminal for children who give no indication of how light gets here.

The network showing children's ideas about sources divides into two inclusive aspects by the use of a 'bra', which are ideas about the nature of the source and the number of ideas mentioned. The latter is necessary because children express or show more than one aspect of the nature of sources so this piece of the network is said to be recursive because it is used more than once. The recursive nature of this part of the network is indicated by the curled arrow (Fig 7.1(c)).

The network for the nature of sources subdivides into an exclusive bar. This is because the data item being looked at represents either a primary or a secondary source but not both. Further subdivisions are then made about primary sources which are exclusive. The subdivisions for the secondary sources are inclusive (shown by the use of a 'bra') and reflect aspects of the children's response.

The ends of the networks are known as terminals. Hence in Table 7.1, the category, 'bulbs' and the category 'pylons' are both examples of terminals. The increasing layers of sub-division within the network are referred to as the increasing 'delicacy' of the network. Such networks are an instrument for data analysis and reflect the view of the researchers. For example, the division between primary and secondary sources is not made by the children and simply provides one perspective for viewing the data.

Each response is then coded. For example, the response shown in Fig 5.1, would require a tick in the following terminals: Bulbs, lights, torches, moon, mirror, fire. If the child, then explained how the mirror was a source of light by saying it was able to reflect light, the terminal 'correctly explained' would have been ticked, otherwise a tick would be placed in the terminal 'no statement'.

# Table 7.1: Network analysis for children's ideas of sources of light showing results for the intervention

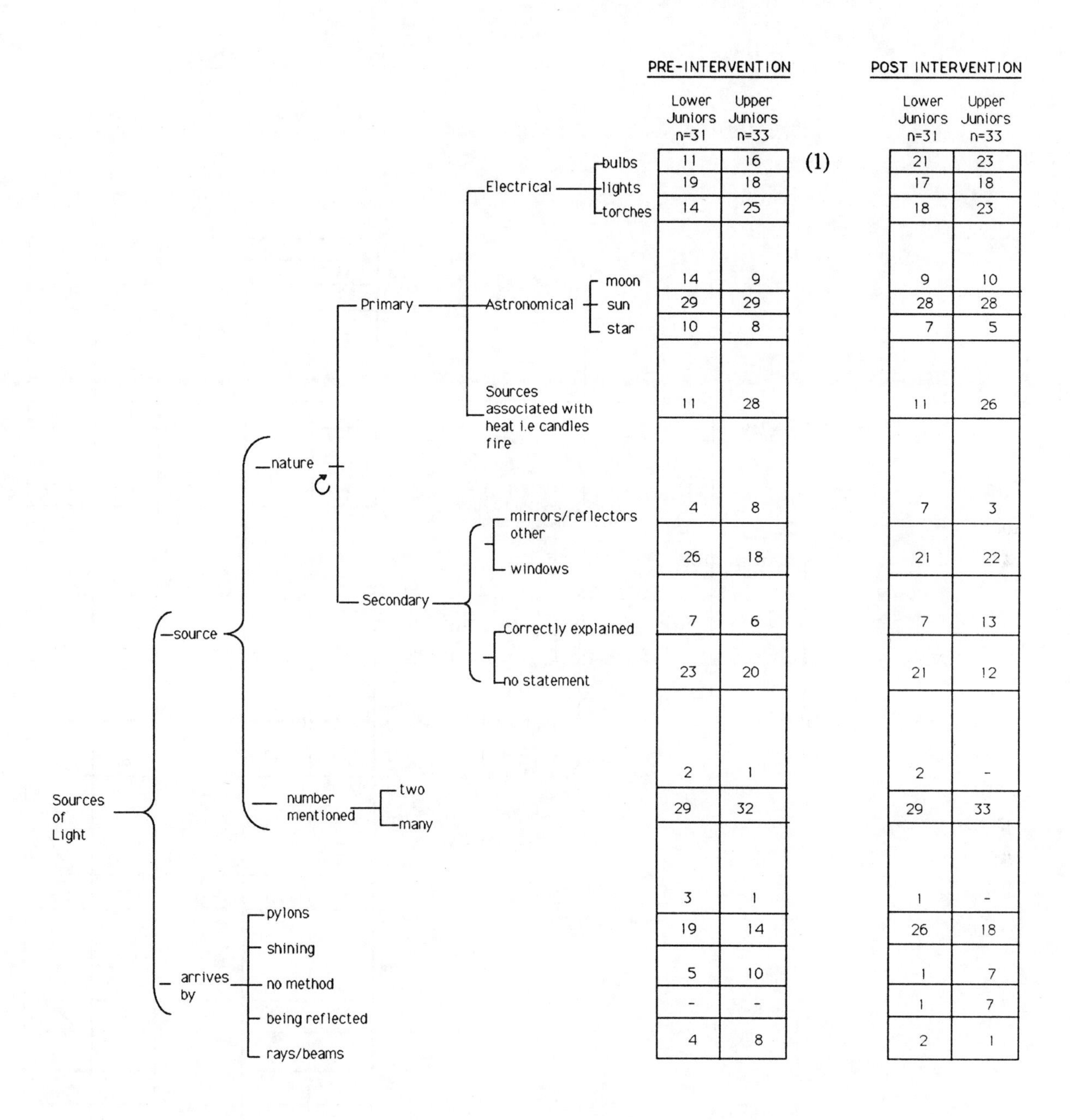

| Sources of Light | Pre-intervention: Lower Juniors n=31 | Pre-intervention: Upper Juniors n=33 | Post intervention: Lower Juniors n=31 | Post intervention: Upper Juniors n=33 |
|---|---|---|---|---|
| source – nature – Primary – Electrical – bulbs (1) | 11 | 16 | 21 | 23 |
| source – nature – Primary – Electrical – lights | 19 | 18 | 17 | 18 |
| source – nature – Primary – Electrical – torches | 14 | 25 | 18 | 23 |
| source – nature – Primary – Astronomical – moon | 14 | 9 | 9 | 10 |
| source – nature – Primary – Astronomical – sun | 29 | 29 | 28 | 28 |
| source – nature – Primary – Astronomical – star | 10 | 8 | 7 | 5 |
| source – nature – Primary – Sources associated with heat i.e candles fire | 11 | 28 | 11 | 26 |
| source – nature – Secondary – mirrors/reflectors other | 4 | 8 | 7 | 3 |
| source – nature – Secondary – windows | 26 | 18 | 21 | 22 |
| source – nature – Secondary – Correctly explained | 7 | 6 | 7 | 13 |
| source – nature – Secondary – no statement | 23 | 20 | 21 | 12 |
| source – number mentioned – two | 2 | 1 | 2 | - |
| source – number mentioned – many | 29 | 32 | 29 | 33 |
| arrives by – pylons | 3 | 1 | 1 | - |
| arrives by – shining | 19 | 14 | 26 | 18 |
| arrives by – no method | 5 | 10 | 1 | 7 |
| arrives by – being reflected | - | - | 1 | 7 |
| arrives by – rays/beams | 4 | 8 | 2 | 1 |

[1] Figures in Table 7.1 show the total number of instances for each terminal of the network

## Table 7.2. Network analysis for children's ideas of sources showing figures for totals.

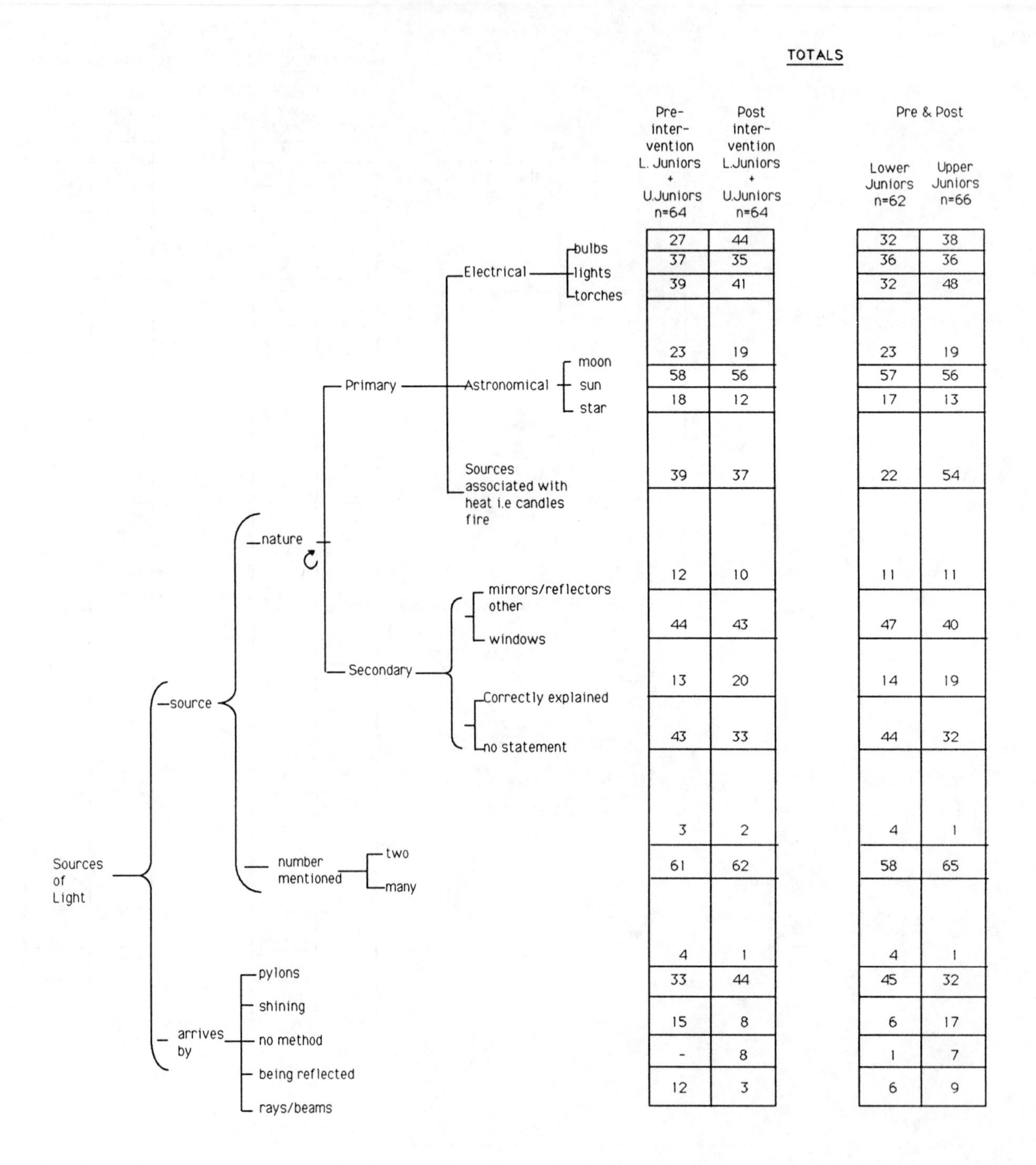


| | Pre-intervention L. Juniors + U.Juniors n=64 | Post intervention L.Juniors + U.Juniors n=64 | Pre & Post Lower Juniors n=62 | Pre & Post Upper Juniors n=66 |
|---|---|---|---|---|
| bulbs | 27 | 44 | 32 | 38 |
| lights | 37 | 35 | 36 | 36 |
| torches | 39 | 41 | 32 | 48 |
| moon | 23 | 19 | 23 | 19 |
| sun | 58 | 56 | 57 | 56 |
| star | 18 | 12 | 17 | 13 |
| Sources associated with heat i.e candles fire | 39 | 37 | 22 | 54 |
| mirrors/reflectors other | 12 | 10 | 11 | 11 |
| windows | 44 | 43 | 47 | 40 |
| Correctly explained | 13 | 20 | 14 | 19 |
| no statement | 43 | 33 | 44 | 32 |
| two | 3 | 2 | 4 | 1 |
| many | 61 | 62 | 58 | 65 |
| pylons | 4 | 1 | 4 | 1 |
| shining | 33 | 44 | 45 | 32 |
| no method | 15 | 8 | 6 | 17 |
| being reflected | - | 8 | 1 | 7 |
| rays/beams | 12 | 3 | 6 | 9 |

Figures in Table 2 show

a. Total for all children pre- and post-intervention.
b. Figures totalled for lower juniors (pre and post-intervention) and upper juniors (pre- and post-intervention).

This network shows two essential features of children's understanding of light sources. Firstly that the sources can be distinguished into primary and secondary sources. Asked to draw 'anything that gives off light', the children predominantly drew primary light sources. Asked 'where is light coming from', children provided responses that show an awareness of secondary sources such as window, mirrors and the ceiling. Only rarely did they offer an explanation of where the light for the secondary source originates i.e 'The light is coming through the windows from the sun.' Statements about secondary sources were normally limited to 'from the windows'.

The second feature of the network is a summary of the responses that children gave to the question 'How does light get here?'. The predominant response here was that objects 'shine' or 'shine down' with no overt recognition of something which travels. The use of the term 'shines' implies a causal recognition of the source of the light i.e the sun causes the light by shining rather than a recognition of a mechanism of transference.

The figures in Table 7.1 and 7.2 are the actual numerical values and show some interesting features. Firstly, nearly all pupils are aware of a wide variety of sources of light. The sources indicated were predominantly primary sources as this summary shows

**Table 7.3. Total Number of Primary and Secondary Sources indicated**

| | *Pre-Intervention* | | *Post-Intervention* | |
|---|---|---|---|---|
| | Lower Juniors (n=31) | Upper Juniors (n=33) | Lower Juniors (n=31) | Upper Juniors (n=33) |
| Number of instances primary sources shown | 108 | 133 | 111 | 134 |
| Mean number of primary sources shown per individual | 3.5 | 4.0 | 3.6 | 4.0 |
| Number of instances secondary sources shown | 30 | 26 | 28 | 25 |
| Mean number of secondary sources shown per individual | 1.0 | 0.8 | 1.0 | 0.8 |

These summary figures indicate there was remarkably little variation in the mean number of sources of light indicated by children before and after the intervention. In addition, the average figures show that children found it easy to indicate a reasonable number of objects which are sources of light and there is little difference between lower and upper juniors. Upper juniors did volunteer more sources but the difference was small and the intervention produced no significant change in this.

Primary sources were mentioned by children 3-4 times more often than secondary sources. However, an examination of Table 7.1 and 7.2 shows that the most common sources mentioned by children was the sun, which was mentioned by a minimum of 85% for any one sample. Other common sources mentioned were torches (minimum 45% of sample) and windows (minimum 55% of sample).

Some further insights can be gained by examining the totals for variation between the elicitation prior to the intervention and post intervention and any variation between responses obtained from totalling the responses from lower juniors and upper juniors. Analysis of the data from this perspective provides and indication of any significant differences that occur between these two groups regardless of the intervention. Table 7.2 shows the figures obtained by totalling the scores in this manner. The figures were tested for statistical significance by the various groupings.

**Table 7.4: Statistical significance of changes.**

| | Significant change in the elicitations after the intervention by: | | | *Overall differences between Lower Juniors & Upper Juniors* |
|---|---|---|---|---|
| | Lower Juniors | Upper Juniors | Total | |
| bulbs shown as sources | $p< 0.05$ | - | $p<0.01$ | - |
| torches shown as sources | - | - | - | $p< 0.05$ |
| heat sources | - | - | - | $p< 0.01$ |
| no statement about secondary sources | - | ($p< 0.05$)[1] | - | ($p< 0.05$) |
| Light arrives by shining | $p< 0.05$ | - | $p< 0.05$ | ($p< 0.01$) |
| No method | - | - | - | $p< 0.05$ |

1 Figures shown in brackets represent significant *decreases*.

Although table 7.4 shows that some of the changes were significant, it is notable that there are more significant differences associated with the change in age range than the intervention. The important point is that for a large number of categories for the network, *there has been no significant change.* One explanation would be that children's ideas about sources of light are well developed and rooted in commonplace observations of light coming from a wide range of primary sources. This would account for the preponderance of primary sources mentioned. Everyday observations do not recognize secondary sources or their nature which would possibly explain why statements about the source of light for mirrors and windows were relatively rare in both groups.

The positive effects of the intervention were very limited. This was not surprising as the preliminary data had already shown that children were familiar with a wide range of sources and it was felt that there was little that could be done to increase their awareness in the time available. Consequently the intervention phase did not primarily address this area of understanding. It is promising that more children can provide some explanation of secondary sources and talk about light 'shining' but given the small numbers, it is best to be sceptical about placing much emphasis on this result.

More interesting was the difference between responses from lower juniors and upper juniors. There were more natural significant differences found between these two groups than as any consequence of the intervention. Apart from fewer upper juniors who explained the arrival of light by shining, they were all positively weighted changes towards a more elaborate model of sources and how light travels. This would suggest that there is some experiential development with age, though it is important to note again that for the majority of categories, there is no significant change.

*Summary:* *The evidence can be summarised as follows.*

a. *Young children show an awareness of a wide variety of sources of light. The sources shown are predominantly primary sources.*

b. *There is some evidence that older juniors have a more complex model of sources which incorporates a recognition of secondary sources of light. Most of this difference can be explained by experiential developmental change rather than any effect of this study.*

c. *The most noticeable feature shown by the data is that there is very little change in children's understanding of sources of light as a result of this intervention.*

### *7.1b Representations of Light*

Many of the elicitation activities called on children to use drawings to provide an explanation of what was happening in the activity or how they achieved a set task. The most notable feature about these tasks was the wide variation in the representations used by children to show what was occurring. These activities were, showing how they were able to see a torch in a mirror; showing how they saw the light from a candle; explaining how they saw a book to their younger brother/sister and showing how they saw a clock on the wall. Children were encouraged to use drawings in their explanations because this was found to be a productive method of obtaining answers from children about their ideas through a familiar mode of expression.

*In this analysis the data about representations has been taken from drawings and explanations which show or discuss light alone.* Representations of vision or links between the eye and object were considered indicative of some understanding of vision and not used for this analysis of children's representations of light.

Again the results obtained have been categorised using a network shown in Table 7.5 & 7.6. These summarise the main features of the representations employed by children. The dominant feature of children's work was the use of lines as a means of representing light from a relatively early age (Fig 6.20 & 6.22). And secondly, to incorporate arrows which showed a sense of direction. It was also noticeable that nearly all children's work included small, short lines around sources (Fig 6.2). The strength of this feature (87% minimum in any one sample) is perhaps surprising and it may be an *a priori* construct to developing a more sophisticated representation.

However, this was not the only representation found. Others were particles where the light was shown as string of small balls or a broken line; a 'sea of light' where the light was shown in shading across the whole drawing; beams where the light was indicated as a broad beam of light rather than a narrow line and 'blobs'. 'Blobs' was the term used by some of the children to describe a patch of light which they draw at the end of the torch or on a mirror or piece of paper.

(contd overleaf)

## Table 7.5. Network analysis of children's representations of light

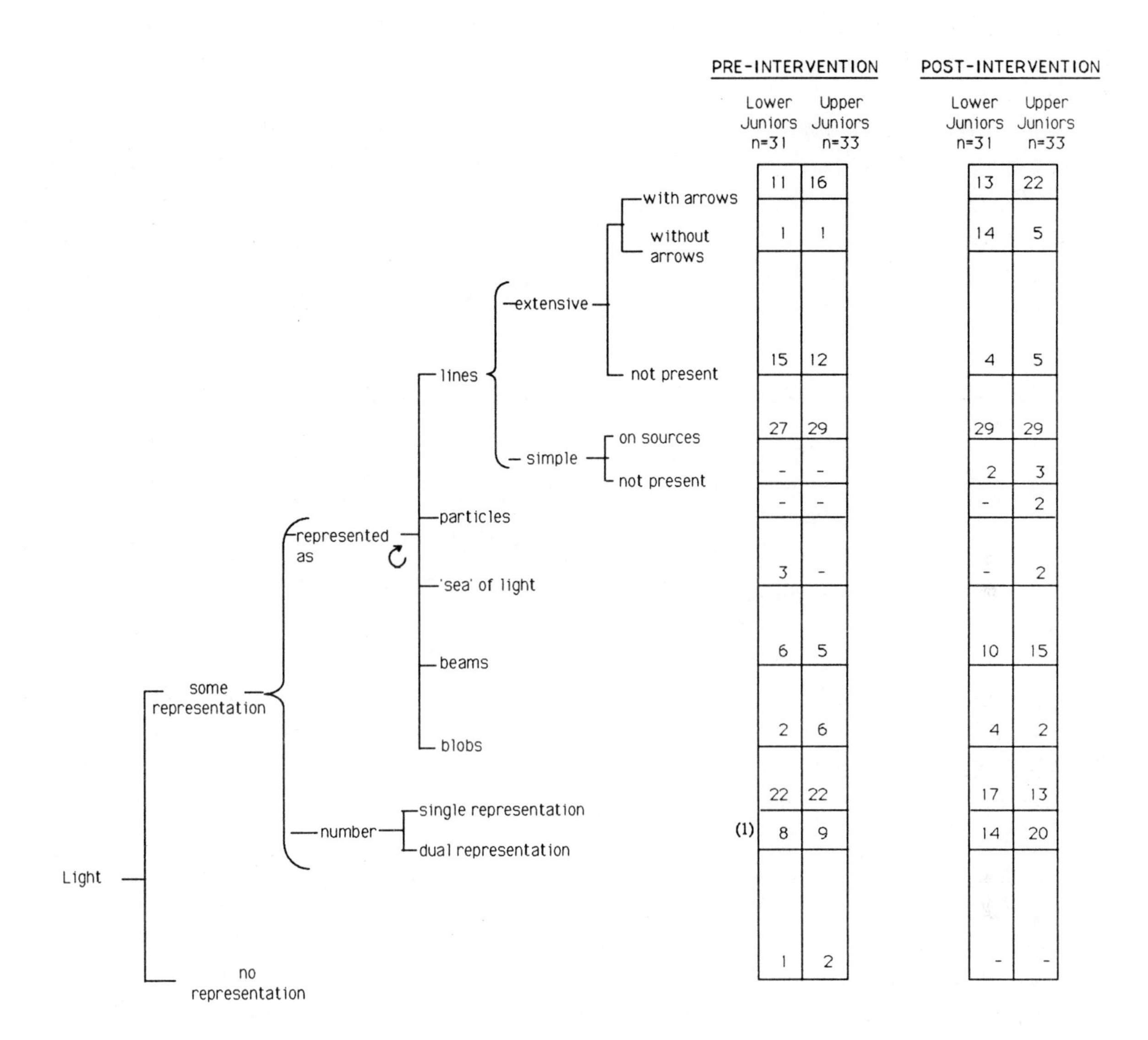

| Light | | | PRE-INTERVENTION Lower Juniors n=31 | PRE-INTERVENTION Upper Juniors n=33 | POST-INTERVENTION Lower Juniors n=31 | POST-INTERVENTION Upper Juniors n=33 |
|---|---|---|---|---|---|---|
| some representation – represented as – lines | extensive | with arrows | 11 | 16 | 13 | 22 |
| | | without arrows | 1 | 1 | 14 | 5 |
| | | not present | 15 | 12 | 4 | 5 |
| | simple | on sources | 27 | 29 | 29 | 29 |
| | | not present | - | - | 2 | 3 |
| some representation – represented as | particles | | - | - | - | 2 |
| | 'sea' of light | | 3 | - | - | 2 |
| | beams | | 6 | 5 | 10 | 15 |
| | blobs | | 2 | 6 | 4 | 2 |
| some representation – number | single representation | | 22 | 22 | 17 | 13 |
| | dual representation (1) | | 8 | 9 | 14 | 20 |
| no representation | | | 1 | 2 | - | - |

1 These figures show the total No of representations. In this example 22 lower juniors use a single representation and 8 use a dual representation which makes a total of 38 representations. The nature of these representations is shown by the upper half of the network.

## Table 7.6. Network analysis of children's representations of light. Data for Totals

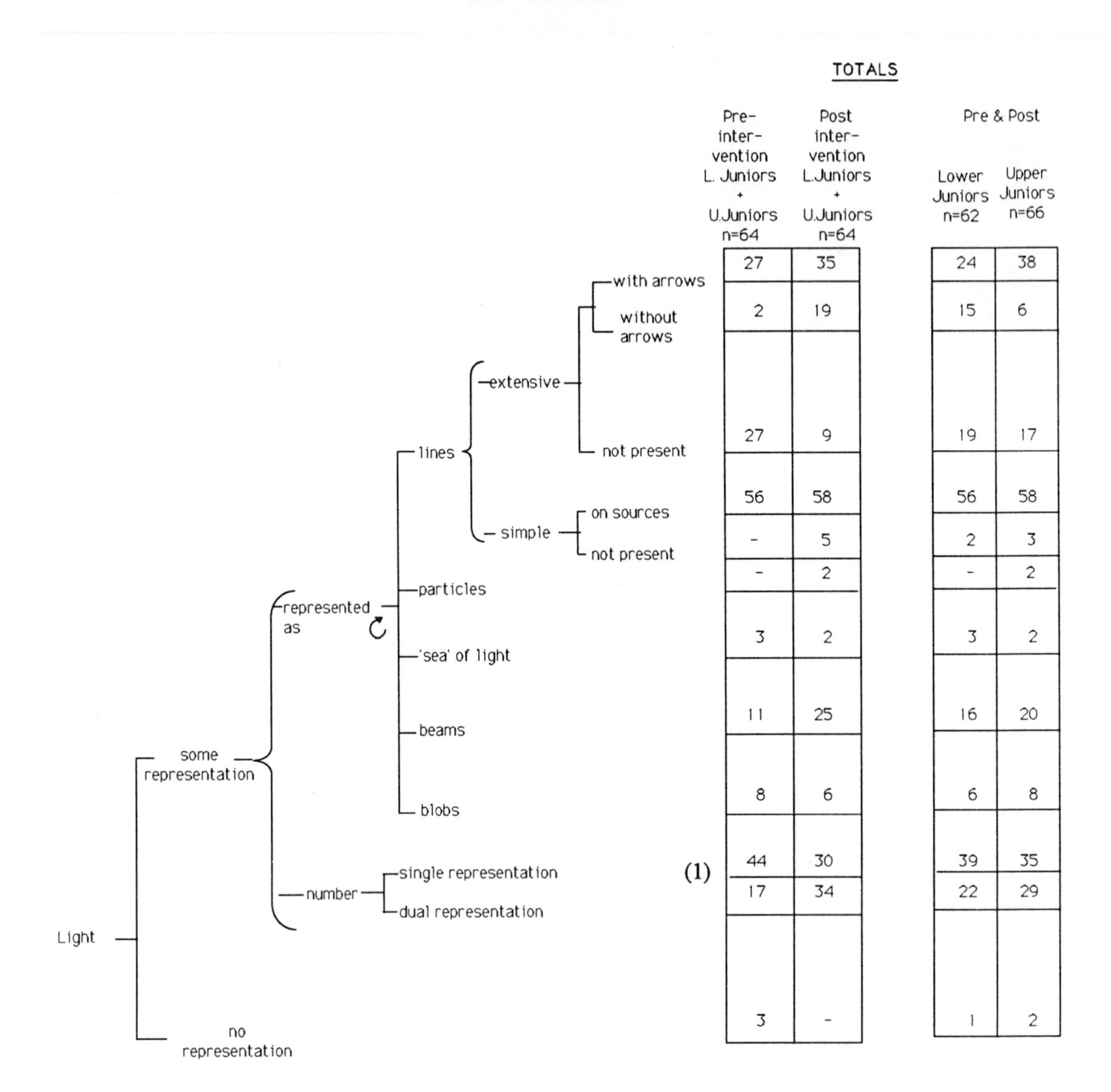

| Light | | | | | TOTALS: Pre-intervention L. Juniors + U.Juniors n=64 | TOTALS: Post intervention L.Juniors + U.Juniors n=64 | Pre & Post: Lower Juniors n=62 | Pre & Post: Upper Juniors n=66 |
|---|---|---|---|---|---|---|---|---|
| some representation | represented as | lines | extensive | with arrows | 27 | 35 | 24 | 38 |
| | | | | without arrows | 2 | 19 | 15 | 6 |
| | | | | not present | 27 | 9 | 19 | 17 |
| | | | simple | on sources | 56 | 58 | 56 | 58 |
| | | | | not present | - | 5 | 2 | 3 |
| | | particles | | | - | 2 | - | 2 |
| | | 'sea' of light | | | 3 | 2 | 3 | 2 |
| | | beams | | | 11 | 25 | 16 | 20 |
| | | blobs | | | 8 | 6 | 6 | 8 |
| | number (1) | single representation | | | 44 | 30 | 39 | 35 |
| | | dual representation | | | 17 | 34 | 22 | 29 |
| no representation | | | | | 3 | - | 1 | 2 |

[1] See the note on Table 7.5

The range of the representations used is surprising. Some representations were possibly rooted in observations of beams of light from torches and 'blobs' on paper. However, the observational evidence for light consisting of lines is relatively tenuous and based on limited observations. Little evidence was found here for the notion that humans and objects exist in a 'sea of light'. However, this may reflect a failure of the elicitation and a difficulty for children in representing such a concept.

The other feature of the children's work is the appearance of representations which are context dependent and vary with the task. Though the majority used a single representation, there were a number who would switch the representation from one question to the next. Time did not allow an investigation of whether those children who showed a single representation were being consistent or whether the elicitation failed to evoke another representation. One interpretation is that many representations are observationally dependent and that the children just drew what they observe i.e beams with torches and lines with candles.

The figures are presented in a similar manner to tables 7.1 and 7.2. There are essentially three main features to the tables that should be noted. Firstly, the intervention has produced a significant increase in the number of children using lines as a representation for light.

**Table 7.7. Total Number of Children Using Extensive Lines to Represent Light**

| | *Pre-Intervention* | | *Post-Intervention* | |
|---|---|---|---|---|
| | Lower Juniors n=31 | Upper Juniors n=33 | Lower Juniors n=31 | Upper Juniors n=33 |
| Extensive Lines | 12 | 17 | 27 | 27 |
| Percentage | (39) | (52) | (87) | (82) |
| Totals | 29 | | 54 | |

Table 7.7 shows the values obtained by adding the first two terminals on the network together. The result shows a significant change ($p<0.01$) in the number of children using lines to represent light for both groups and the totals. A closer examination of the network shows that the changes for lower juniors can be explained by a larger number who used lines showing no sense of direction. For upper juniors, the significant change is due to a larger number of children who showed representations of light with a sense of direction.

However, it is notable that the totals for lower juniors and upper juniors show a significant increase ($p<0.05$) in the number who used arrowed links anyway. If the ability to represent light in the form of a line is considered indicative of a more sophisticated model, an implication of this result is that children of age 9-11 are developing the ability to think with such models anyway. However, the increase in significance suggests that the intervention, with its emphasis on drawing and representing light, may have contributed to this development.

The second feature of the network was the increase in the number of upper juniors using beams as a means of representing light. All the other significant changes occurred for upper juniors and these are shown in Table 7.8.

**Table 7.8. Statistical significance of changes for representations**

| | *Lower Juniors* | *Upper Juniors* | *Totals* |
|---|---|---|---|
| Representations as beams | - | $p<0.01$ | - |
| Single Representations | - | ($p<0.05$) | - |
| Dual Representations | - | $p<0.01$ | $p<0.05$ |

The increase in the number of children who used beams to represent light has no clear explanation other than that it may be based in more careful and thorough observation of car headlamps and torches. What the figures do suggest is that more children are using beams *and another* representation for light so that there is a decrease in the single representations and an increase in the dual representations which is the third feature of the networks. Possibly, this is indicative of a greater fluidity in children's understanding which although richer in its repertoire, is still very context-specific. Finally it is worth noting that very few children provide no representation of light in their responses.

*Summary:* *The evidence can be summarised as follows.*

a. *Nearly all children will represent light around a source with short lines.*

b. *The majority of upper junior children showed light using extensive lines. The representation of light as a ray or line was seen to increase between the ages of 7 and 11. Part of this development would appear to occur with age and some of the development could be explained as a consequence of the specific intervention activities.*

c. *Representations of light used by upper junior children become more varied and context dependent. Significantly more children provided responses that used more than one representation of light to answer similar questions after the intervention. Part of the increase could be explained by a significant change in the number that use beams to represent light.*

d. *Nearly all children provided some representation of light.*

### *7.1c Young children's Understanding of the Nature of Vision*

Three topics in the elicitation materials addressed the nature of vision and the understanding shown by children. Pupils were asked (a) to show how they were able to see the light from a torch in a mirror, (b) to explain how they saw a book to a younger brother and (c), to add to a drawing to show how they saw a clock on the wall. These activities produced a wide range of responses which are summarised in Table 7.9 and 7.10 and the main features are discussed here.

**Table 7.9. Network Analysis of children's responses to questions about the Nature of Seeing.**

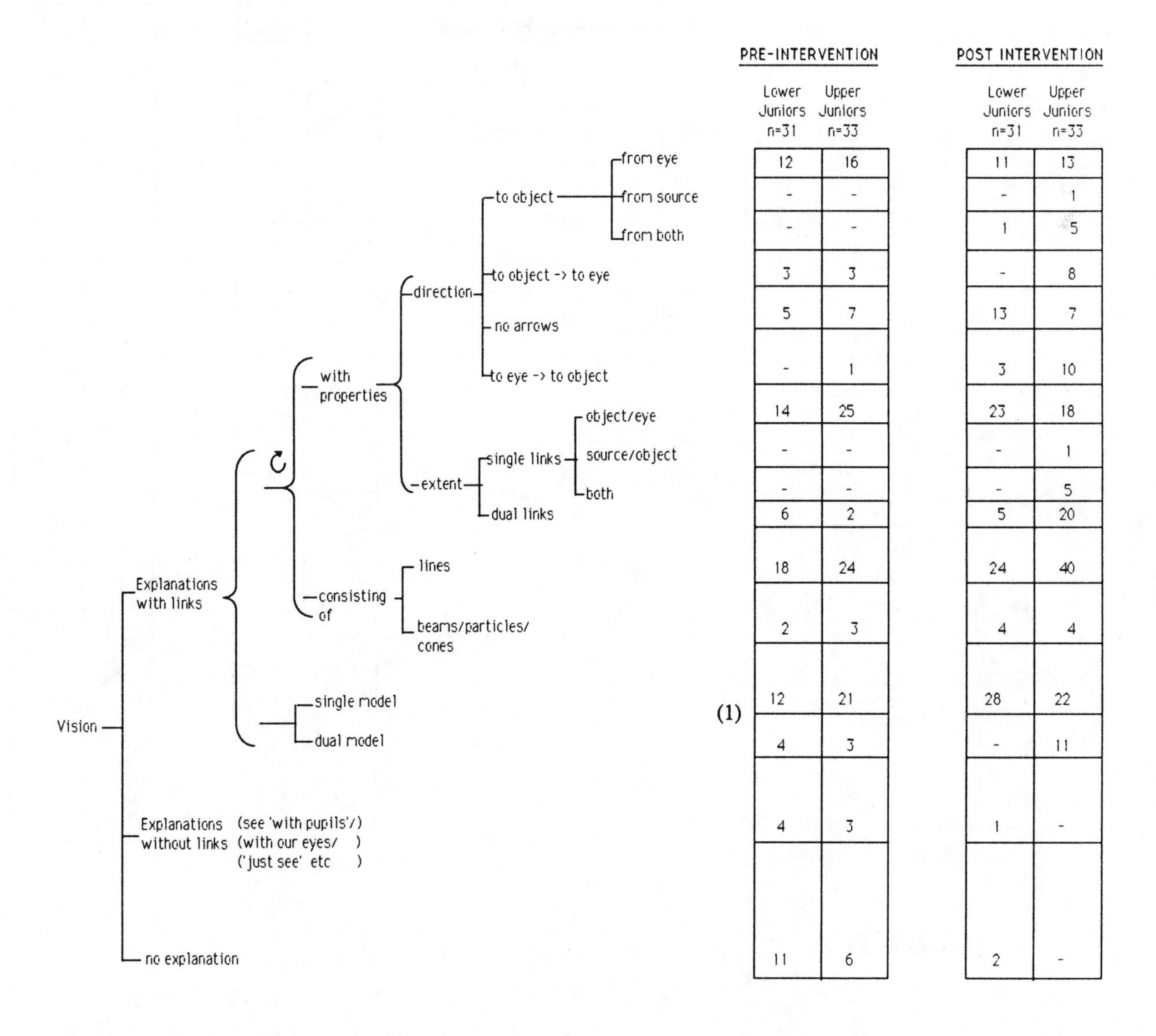

| Vision | PRE-INTERVENTION Lower Juniors n=31 | PRE-INTERVENTION Upper Juniors n=33 | POST INTERVENTION Lower Juniors n=31 | POST INTERVENTION Upper Juniors n=33 |
|---|---|---|---|---|
| Explanations with links – with properties – direction – to object – from eye | 12 | 16 | 11 | 13 |
| – from source | - | - | - | 1 |
| – from both | - | - | 1 | 5 |
| – to object -> to eye | 3 | 3 | - | 8 |
| – no arrows | 5 | 7 | 13 | 7 |
| – to eye -> to object | - | 1 | 3 | 10 |
| – extent – single links – object/eye | 14 | 25 | 23 | 18 |
| – source/object | - | - | - | 1 |
| – both | - | - | - | 5 |
| – dual links | 6 | 2 | 5 | 20 |
| – consisting of – lines | 18 | 24 | 24 | 40 |
| – beams/particles/cones | 2 | 3 | 4 | 4 |
| – single model (1) | 12 | 21 | 28 | 22 |
| – dual model | 4 | 3 | - | 11 |
| Explanations without links (see 'with pupils'/) (with our eyes/ ) ('just see' etc ) | 4 | 3 | 1 | - |
| no explanation | 11 | 6 | 2 | - |

Notes

1 The network is best understood by examining the figures for 'single' and 'dual' models. i.e for Lower juniors prior to the intervention, there were 12 single model responses (with links) and 4 dual model responses making a total of 20 responses in all. The upper half of the network shows what form these responses took.

**Table 7.10. Network Analysis of Children's responses about Vision showing data for total figures.**

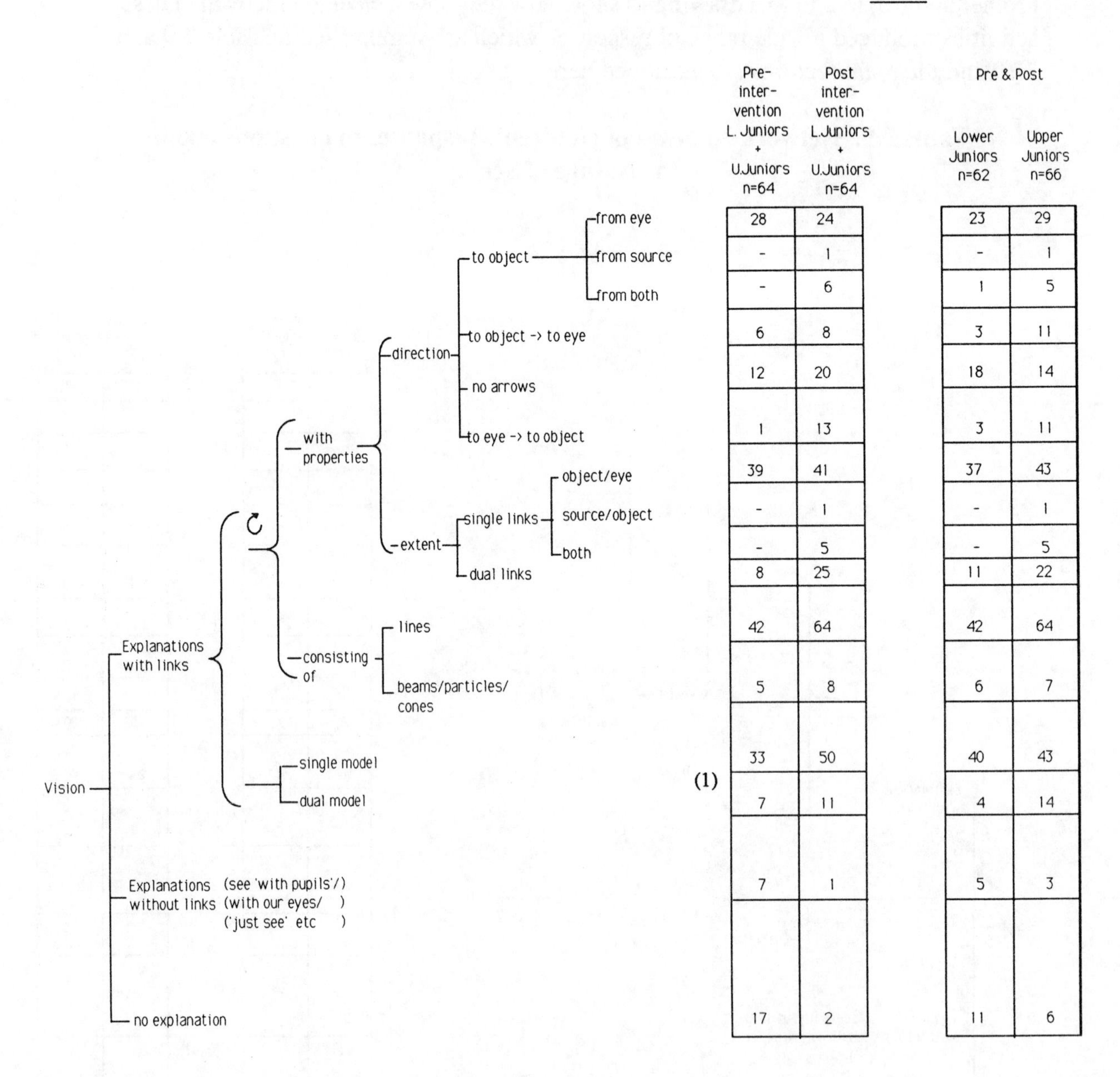

| Vision | Pre-intervention L. Juniors + U.Juniors n=64 | Post intervention L.Juniors + U.Juniors n=64 | TOTALS Pre & Post Lower Juniors n=62 | TOTALS Pre & Post Upper Juniors n=66 |
|---|---|---|---|---|
| Explanations with links — with properties — direction — to object — from eye | 28 | 24 | 23 | 29 |
| direction — to object — from source | - | 1 | - | 1 |
| direction — to object — from both | - | 6 | 1 | 5 |
| direction — to object -> to eye | 6 | 8 | 3 | 11 |
| direction — no arrows | 12 | 20 | 18 | 14 |
| direction — to eye -> to object | 1 | 13 | 3 | 11 |
| extent — single links — object/eye | 39 | 41 | 37 | 43 |
| extent — single links — source/object | - | 1 | - | 1 |
| extent — single links — both | - | 5 | - | 5 |
| extent — dual links | 8 | 25 | 11 | 22 |
| consisting of — lines | 42 | 64 | 42 | 64 |
| consisting of — beams/particles/cones | 5 | 8 | 6 | 7 |
| single model (1) | 33 | 50 | 40 | 43 |
| dual model | 7 | 11 | 4 | 14 |
| Explanations without links (see 'with pupils'/) (with our eyes/ ) ('just see' etc ) | 7 | 1 | 5 | 3 |
| no explanation | 17 | 2 | 11 | 6 |

1 See note 1 on Table 7.9

The data shows that children of all ages readily produced responses that used links between the eye and the object and that many incorporated the sense of direction. The number of such responses is shown in Table 7.11.

**Table 7.11. Analysis of responses obtained which show links between *eye* and *object***

| | *Pre-Intervention* | | *Post-Intervention* | |
|---|---|---|---|---|
| | Lower Juniors n=31 | Upper Juniors n=33 | Lower Juniors n=31 | Upper Juniors n=33 |
| No. of Responses showing link between Eye and Object. | 20 | 27 | 28 | 44 |
| No. of responses showing single links | 14 | 25 | 23 | 24 |
| Responses showing Dual Links | 6 | 2 | 5 | 20 |

The data shows a significant increase in the number of both upper and lower juniors who show a link between eye and object. The other noticeable feature of table 7.11 is the decline in the number of responses from upper juniors to explain vision using single links between *the eye* and *object.* These responses dropped from 25 out of 27 to 24 out of 44 which is significant ($p < 0.01$). This was accompanied by a significant increase in responses from upper juniors which showed dual links to 20 out of 44 responses ($p < 0.01$).

Table 7.12 shows that for those that showed a link between eye and object, the majority of children at all ages incorporated a sense of direction into their response about vision. None of these changes are significant.

**Table 7.12. Percentage of responses showing the sense of direction of vision.**

| | *Pre-Intervention* | | *Post-Intervention* | |
|---|---|---|---|---|
| | Lower Juniors n=31 | Upper Juniors n=33 | Lower Juniors n=31 | Upper Juniors n=33 |
| Arrowed (%) Link | 75 | 69 | 54 | 84 |

Another noteworthy point is that there was a small minority of children in the lower juniors (15%) who are able to provide responses in terms of accepted scientific theories of vision by indicating that the light goes to the object and then to the eye. However, it would appear that such thinking was not robust as the post intervention data showed that no lower junior children had this model. Lower juniors also showed a large minority (35%) of children who provided no explanation for vision.

One weakness of the network is a failure to show children who indicated in writing in one context that vision occurs 'with our eyes' or that we 'just see the book' in addition to providing a drawing as another response. The number of such responses was counted separately and shown in Table 7.13.

**Table 7.13. Percentage of children providing additional written responses to explain vision.**

| | *Pre-Intervention* | | *Post-Intervention* | |
|---|---|---|---|---|
| | Lower Juniors n=31 | Upper Juniors n=33 | Lower Juniors n=31 | Upper Juniors n=33 |
| Written (%) response | 29 | 24 | 55 | 24 |

The change for lower juniors was just significant ($p<0.05$) but there is no evidence to explain this change. The data indicates that there are a number of children who view vision in certain contexts as being essentially non-problematic. Seeing is just something which happens and you see with your eyes. However, the networks show that there is a very small number of children who use this response solely. For the lower juniors, there was also a minority who offered no meaningful response to explain vision.

An analysis of the responses which showed statistically significant changes is summarised in Table 7.14. In this network it is possible for a child to appear in any one of the upper terminals twice, depending upon the responses that they provide. The significance of changes has been evaluated by considering the change in the total number of responses of any one type in relation to the total number of responses. For instance the number of responses from upper juniors, which show vision in terms of a single link to object from the eye, decreases from 16 out of 27 responses to 13 out of 44 responses.

Table 7.14 shows that the majority of changes have occurred for the upper juniors. These can be summarised as a decrease in the number of children using responses which showed a link from the eye to the object; a decrease in single links; a decrease in responses without links and a decrease in responses which provided no explanation. This was coupled with an increase in the number that showed an explanation with dual links and used dual models to explain vision. However, the latter is not accounted for by an increase in the number of children using scientific models of vision but by a growth in the number of children using explanations that show the light going to the eye and then to the object. This result suggests that more children were aware that 'light is necessary for vision' and 'eyes are needed to see' and were attempting to show both features.

**Table 7.14: Statistical significance of changes in children's ideas about vision.**

| | Significant change in the elicitation after the intervention by: | | | *Overall differences between Lower Juniors & Upper Juniors* |
|---|---|---|---|---|
| | Lower Juniors | Upper Juniors | Total | |
| light shown to object from eye | - | ($p<0.01$)[1] | ($p<0.01$) | - |
| no arrows shown | - | - | - | ($p<0.05$) |
| light shown to eye to object | - | $p<0.05$ | $p<0.01$ | - |
| single links shown object-eye | - | ($p<0.01$) | ($p<0.01$ | - |
| dual links | - | $p<0.01$ | $p<0.05$ | - |
| single models of vision | $p<0.01$ | - | $p<0.01$ | - |
| dual models of vision | - | $p<0.05$ | - | $p<0.05$ |
| Explan-ations without links | - | ($p<0.05$) | ($p<0.05$) | - |
| no explan-ation | ($p<0.05$) | ($p<0.05$) | ($p<0.01$) | - |

[1] Changes representing decreases are shown in brackets

Part of the increase in the use of dual models can be explained by the change observed between lower juniors and upper juniors which is significant. The only other significant changes between the lower junior cohort and the upper junior cohort was a reduction in the number of children who showed no arrows on their drawings. This suggests that part of the observed change in the use of dual models can explained by development which occurs with age.

The increase in dual models would support the hypothesis that children's thinking is context specific and that their ideas are fluid and pliant which has been mentioned elsewhere.

The lower junior children show very few changes. The principal change is an increase in the number of children who show a single representation. This could be explained by the significant decrease in the number of children who provide no meaningful response which suggests that these children are now showing at least one explanation for vision of greater complexity.

Finally, the observed changes are somewhat surprising as the intervention avoided directly addressing this idea of vision. It is possible that the children placed a different emphasis on the activities to that intended. Explanations about the phenomena may be of an egocentric nature which places an emphasis on the function of sight. However, there is no evidence which provides more insight.

In summary, it is clear that the intervention has had more effect on children's development for the upper juniors than lower juniors. This is a similar conclusion to that drawn from looking at the representations for light. The inference is that such work has possibly more value if tackled at a later stage in a junior child's development.

*Summary:* *The evidence can be summarised as follows.*

*a.* *More than half the children provide responses which indicate a link between eye and object and the majority of these responses incorporate a sense of direction.*

*b.* *A sizeable proportion of lower junior children (35%) provide responses which show no explanation of vision and indicate that the idea is non-problematic for them.*

*c.* *The major effect of the intervention work was on upper junior children who provided more responses which showed increased use of dual links i.e eye-object and object-source. This was accompanied by an increase in the number of dual models reflecting an increase in the context dependence of responses. The implication is that such work is more appropriate to children in the 9-11 age range.*

*d.* *The only significant effect of the intervention for lower junior children was to increase the number of responses showing single links between object and eye and decrease those showing responses which provided no explanation.*

*7.1d Context dependence of responses.*

For the purpose of this study, we have used the notion of 'context dependence' to describe the responses of children which show different representations of light or different mechanisms of vision *within the same elicitation*. An example of such a response is shown in Fig 6.26(a) & (b). Throughout the study, this was one of the most noticeable features about the responses obtained from children. Table 7.15 summarises the figures from the network and shows the percentage of children showing such responses.

**Table 7.15: Percentage of children providing responses which show more than one model and which are inconsistent.**

| | *Pre-Intervention* | | *Post-Intervention* | |
|---|---|---|---|---|
| | Lower Juniors n=31 | Upper Juniors n=33 | Lower Juniors n=31 | Upper Juniors n=33 |
| For representations of light | 25% | 27% | 45% | 60% |
| For explanations of vision | 13% | 9% | 0% | 33% |

The changes for upper juniors were significant and showed an increase in the use of context dependent models. Ideally science education should try and facilitate the construction of robust understandings that are generalisable. This was clearly not the case here and it is possible that such a period may be the precursor of the development of ideas which are more permanent and closer to a scientific understanding.

There are various alternative explanations for such behaviour. A Piagetian perspective would be that all these children were exhibiting early or late concrete thinking which is essentially tied to the observable features of such phenomena. Consequently, the children do not perceive any inconsistency in the different representations which would be apparent to a formal thinker. For them, there simply was no conflict.

However, it may simply be a period of trying a new idea whilst clinging to an old interpretation - indeed, perhaps an essential stage in the development of children's thinking. What it indicates is an addition to the child's ideas for making sense of the world. The implications for teaching are that children should have the opportunity to test their thinking by trying such ideas. Only this would provide the necessary experience to develop their understanding.

*Summary:* a. *Many children's responses to questions about their understanding of light showed different answers in different contexts.*

## *8. CHANGES IN INDIVIDUAL CHILDREN*

The other method for examining change is to look at what has happened to individuals. The networks provide a summary of the whole cohort, but are poor at providing insight into any of the changes that occurred for individual children. Such an analysis is important to obtain a picture of how the changes observed in the network arose. Consequently, it was necessary to develop a method of analysis that would provide some insight into any individual change.

The chosen method was based on the fact that an examination of the data shows clearly definable features of children's understanding of light. A child's representations of light can be classified into groupings which can be said to be a) No representations, b) simple links with lines or beams, c) dual links and d) 'blobs'. Similarly an examination of the direction indicated for light and for vision, can be grouped into a) no direction, b) sense of direction indicated and c) accurate sense of direction. Data for changes in children's representations of light are shown in table 8.1 and 8.2 whilst table 8.3 and 8.4 show the data for changes in their ideas about vision. The large number of questions in each activity provided a large data sample for the size of the group.

In the tables, aspects of children's understanding are enclosed in ellipses and the tables show counts for the number of individuals who have changed their representation of light (Table 8.1 and 8.2) between the elicitation activities, pre- and post-intervention for the *same question.* The numbers in boxes within the ellipses, show the counts for those children who did not change the representation that they used.

The charts shown are summaries for three questions in the elicitation activities and tables for each question are provided later. The figures can be summarised by grouping into three categories; (i) those which show no change; (ii) those which show a change to a view which shows a more complex representation of light and explanation for vision; that is, one which is considered to have more of the features of a scientific representation; (iii) those which show a less sophisticated representation. It is clear from these charts that as well developing their understanding, some children go 'backwards' i.e away from a scientific view in their understanding.

(contd overleaf)

**Table 8.1. Conceptual Map of Changes in Children's Representations of Light (Lower Juniors)**

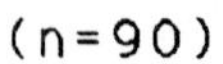

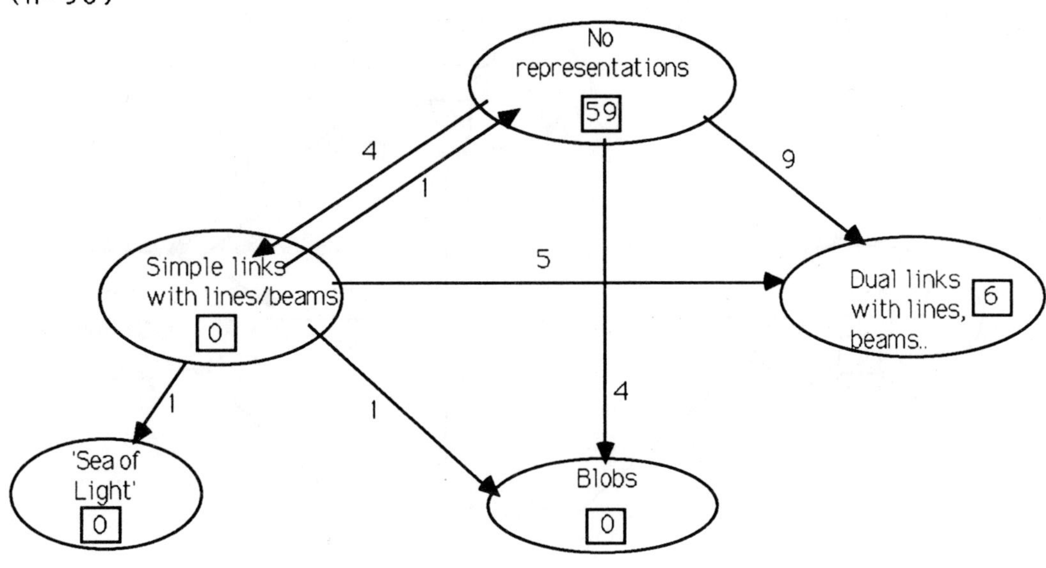


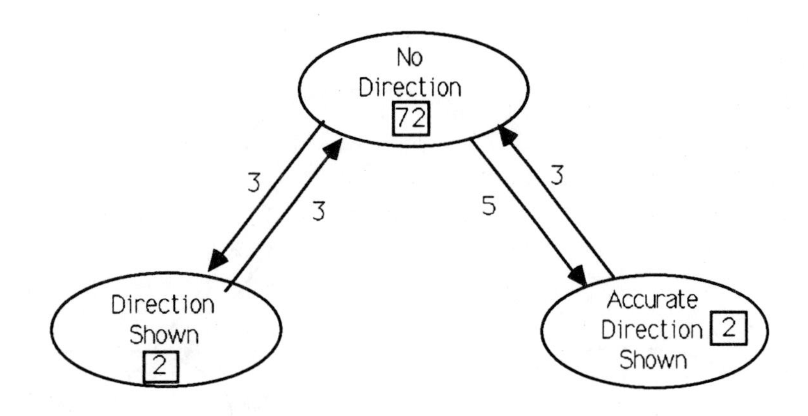

**Table 8.2. Conceptual Map of Changes in Children's Representations of Light (Upper Juniors)**

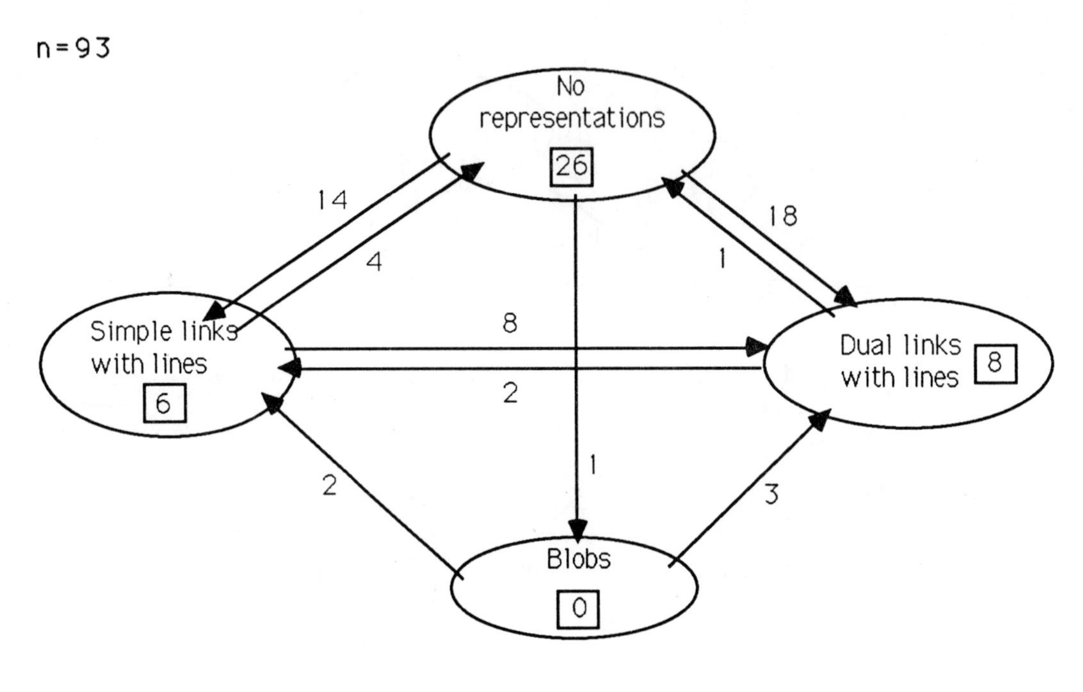


Direction

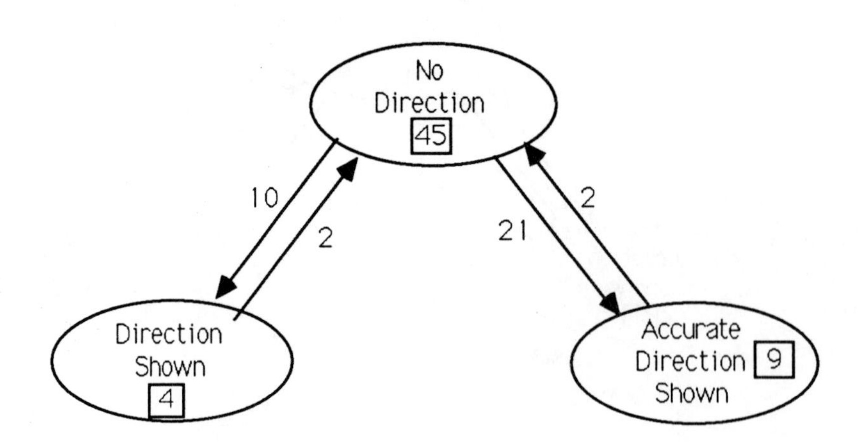

**Table 8.3. Conceptual Map of Changes in Children's Response to Explain Vision (Lower Juniors)**

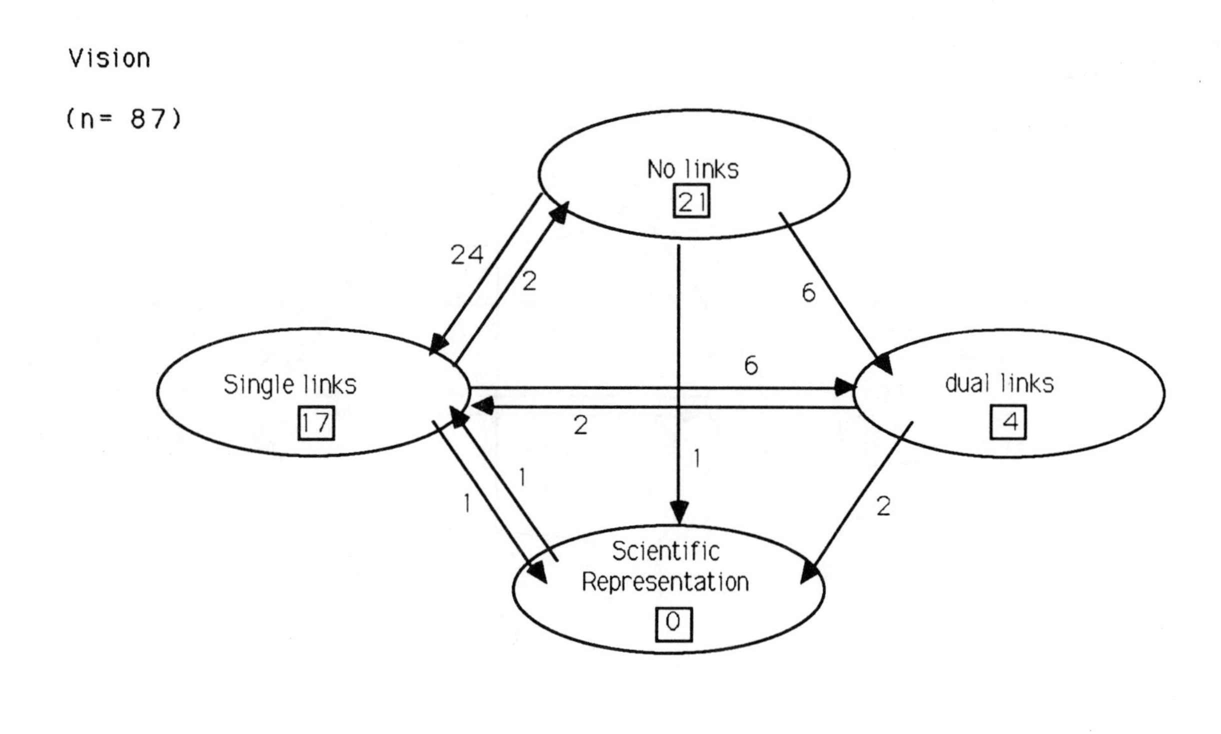


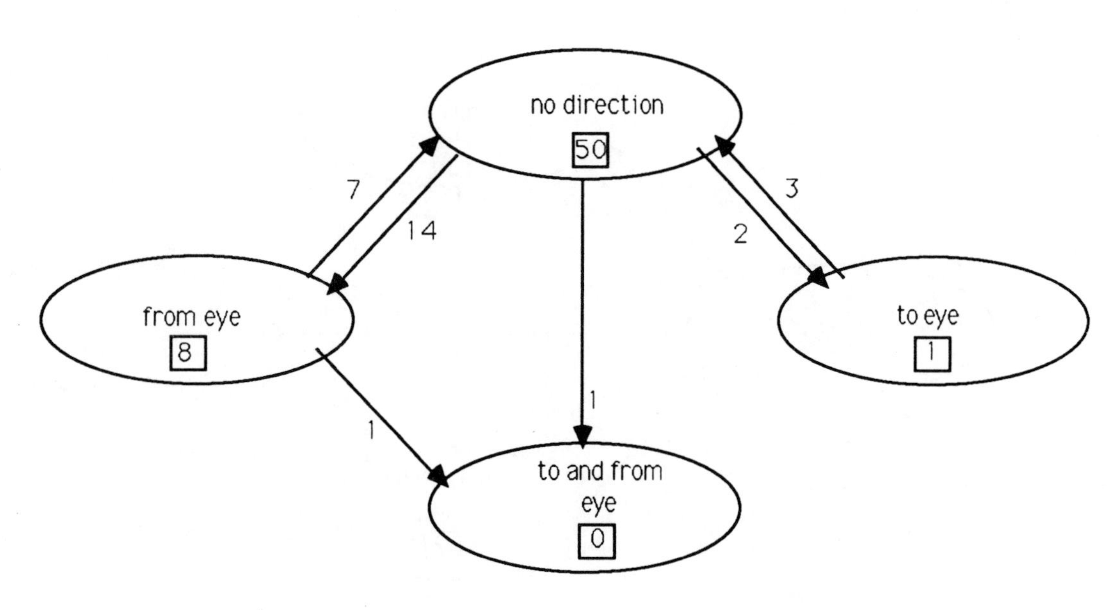

**Table 8.4. Conceptual Map of Changes in Children's Response to Explain Vision (Upper Juniors)**

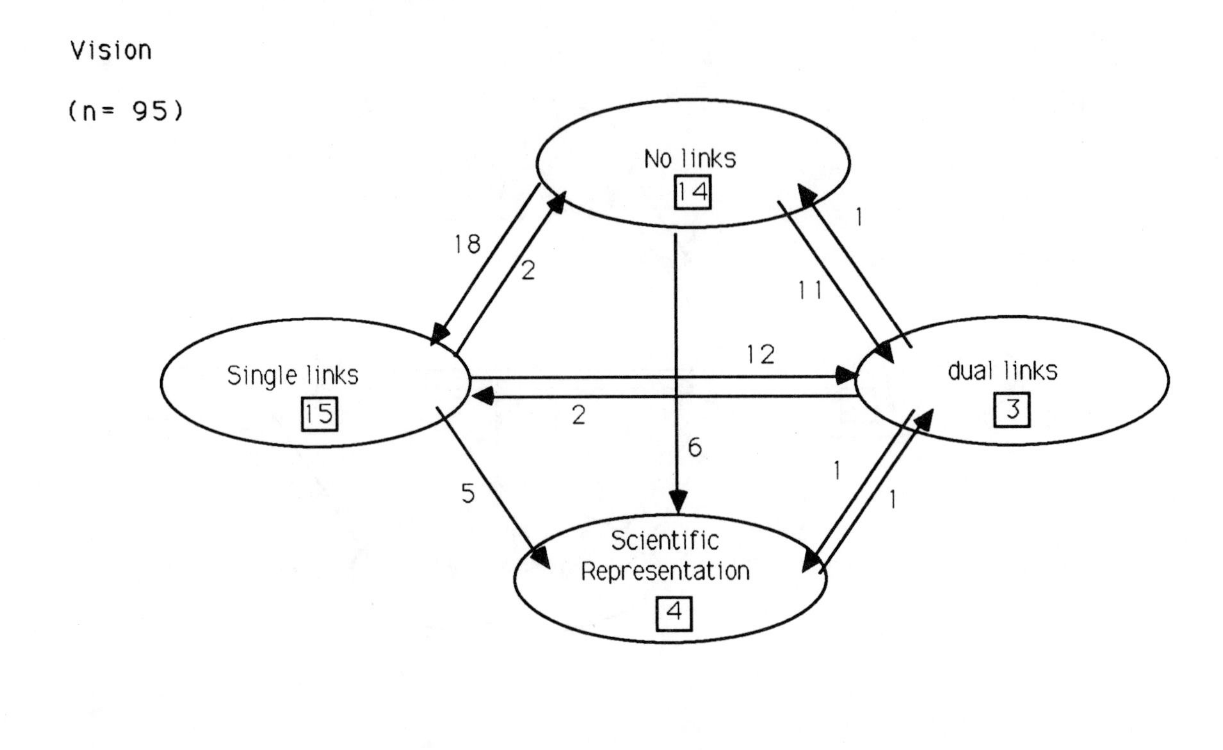


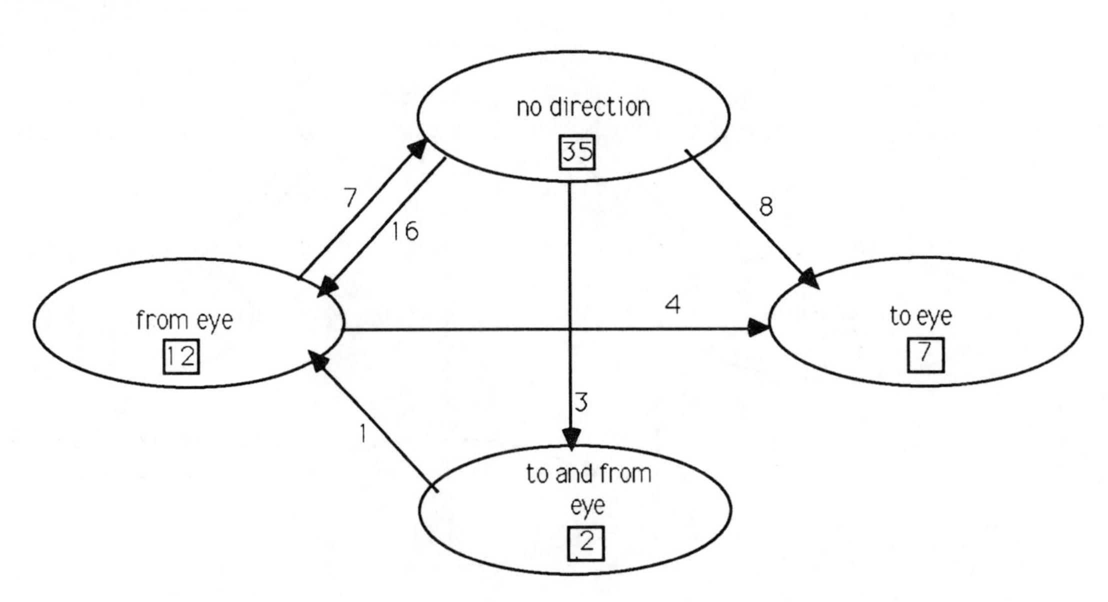

The data in these charts can be summarised in tabular form

**Table 8.5. Summary figures for changes in children's representations of light.**

| | | No Change | Change to more features of a scientific model | Change to less features of a scientific model |
|---|---|---|---|---|
| UPPER | | | | |
| JUNIORS | Representations | 40 | 46 | 7 |
| (n=93) | Direction | 58 | 31 | 4 |
| LOWER | | | | |
| JUNIORS | Representations | 65 | 22 | 3 |
| (n=90) | Directions | 76 | 8 | 6 |

A chi-squared test of the figures in Table 8.5 shows that there was a significant difference ($p<0.01$) for representations and directions between upper and lower juniors. This confirms the analysis of the networks which showed that more change was occurring for the upper juniors than the lower juniors. An analysis of the response for individual items shows that the main contribution to this difference was provided by responses to questions about seeing a book. This would support the notion that it was only upper junior children who are more likely to develop a model of light which is applicable to situations where the source of light is not evident.

The other feature of this analysis is that there were many more children showing no sense of direction i.e no arrows in their drawings of light, both before and after the intervention, compared to those that showed no representation of light. This merely shows that children are often prepared to draw links without indicating a sense of direction. In the case of beams or particle representations, directions were only very rarely indicated.

Table 8.6 shows similar figures for children's ideas about vision and the changes that occurred.

**Table 8.6. Summary figures for changes in children's ideas of vision.**

| | | No Change | Change to more features of a scientific model | Change to less features of of a scientific model |
|---|---|---|---|---|
| UPPER | | | | |
| JUNIORS | Vision | 36 | 53 | 6 |
| (n=95) | Direction | 56 | 31 | 8 |
| LOWER | | | | |
| JUNIORS | Vision | 42 | 40 | 5 |
| (n=87) | Directions | 59 | 18 | 10 |

There was no significant difference between the upper and lower juniors in the changes that have taken place for their ideas of vision or the direction they showed in their explanations. This is perhaps surprising in that it implies that the intervention has had more effect for lower juniors than that indicated by the network analysis.

More insight into the changes is provided by looking at the responses to particular questions using this method of analysis. The figures for the overall results are shown in Table 8.7 & 8.8

**Table 8.7.** **Number of children who showed *NO CHANGE* on questions eliciting representations of light**

| *Question* | *Candle* | *Book* | *Clock* |
|---|---|---|---|
| UPPER JUNIORS | 10(34)[1] | 17(30) | 13(29) |
| Direction indicated | 16 | 22 | 20 |
| LOWER JUNIORS | 13(30) | 29(30) | 23(30) |
| Direction Indicated | 21 | 29 | 26 |

1 Figures in brackets show the number of responses on this item.

These figures show that, for specific questions, there has been little change in the representation elicited from the pupils. This is particularly true of activities associated with secondary sources of light for lower junior pupils. This would suggest that the intervention has had little success in helping children of this age construct a model which represents the role played by light in seeing these objects. There is more success in viewing primary sources where the origin of the light is more tangible. This is confirmed by the figures in Table 8.8 which shows very few lower junior children have moved to a more sophisticated understanding of light for secondary sources.

(contd overleaf)

**Table 8.8.** **Number of children who showed *change* to a more sophisticated representation of light.**

| *Question* | *Candle* | *Book* | *Clock* |
|---|---|---|---|
| UPPER JUNIORS | 21(34) | 13(30) | 12(29) |
| Direction indicated | 14 | 8 | 9 |
| LOWER JUNIORS | 16(30) | 1(30) | 5(30) |
| Direction Indicated | 3 | 1 | 4 |

The figures show that the predominant shift has been for upper juniors and that the largest shift in the response provided, is to a representation showing more features of a scientific understanding for primary sources i.e the candle. This is further evidence that children find it difficult to interpret or explain the phenomena where there is no evident source of light.

A similar analysis of individual responses to elicitation questions about vision was also done.

**Table 8.9** **Number of children who showed *NO CHANGE* on questions eliciting ideas about vision.**

| Question | Candle | Book | Clock |
|---|---|---|---|
| UPPER JUNIORS | 15(35) | 12(31) | 9(29) |
| Direction indicated | 17 | 18 | 21 |
| LOWER JUNIORS | 11(29) | 18(29) | 13(29) |
| Direction Indicated | 18 | 21 | 20 |

**Table 8.10 Number of children who showed *change* on questions eliciting ideas about vision to a more sophisticated understanding.**

| *Question* | *Candle* | *Book* | *Clock* |
|---|---|---|---|
| UPPER JUNIORS | 17(35) | 18(31) | 18(29) |
| Direction indicated | 14 | 10 | 7 |
| LOWER JUNIORS | 15(29) | 9(29) | 16(29) |
| Direction Indicated | 7 | 6 | 5 |

In comparison to the figures for children's representations of light, the data in Table 8.9 & 8.10 reflects that there has been more change in pupil's models of vision and the ideas that they are using as a result of the intervention. They are similar in that the change has been more substantial for upper juniors than for lower juniors.

In all cases the change that has happened has had more effect on the nature of the link than on any sense of direction indicated in the responses. An explanation for this is difficult to provide, other than that the work helped to establish a more concrete representation of the link between light, source and object. Children may not have considered the direction as being something of substantial significance. Many failed to show any sense of direction in their responses to explain vision or of their representation of light.

In conclusion, these data provide a more detailed picture of where changes in children's ideas have occurred. They support the analysis of the networks, in showing that there has been some change, and that has been most significant for upper juniors. It also shows the lack of stability of children's ideas since the picture presented by the data is one of greater overall change than observed in the network analysis. Not only are children developing, but clearly there are some children whose explanations and ideas are 'regressing'. This would support a model of development for children's ideas which is non-linear which may consist of five steps forward and one step back.

(contd overleaf)

*Summary:*

a. *An analysis of the changes in individual children shows that a few children provide dresponses which showed less features of a scientific understanding after the intervention. The overall effect of the intervention has been to develop the understanding of a large number of children whose responses showedmore features of a scientific understanding. However, substantial numbers showed no change in their responses.*

*b. The evidence from these data partially support the analysis provided by the networks, which is that the significant change in understanding has occurred for upper juniors.*

*c. There is noticeably less development in the representations and responses to explain phenomena associated with secondary sources such as a book and clocks. Very few children showed much development here. This would indicate that understanding how we see such objects is an area of conceptual difficulty for primary children.*

## 9. SUMMARY

The following is a summary of the main findings described in section 7 and 8 and provides a resumé of the findings of this phase of the research. The data were obtained by an elicitation phase with children in the classroom from a set of practical experiences. This was known as the pre-intervention period. This was followed by an intervention phase when the children were allowed to try out activities and investigations related to the topic. Following this another set of data were obtained from the children using similar elicitation activities which is referred to here as the post-intervention phase.

The main areas of note were found to be:

### *9.1 Children's Understanding of Sources of Light.*

All ages of children were able to show a knowledge of a wide range of sources and there was no evidence for any change as a result of the intervention or through development with age. However, it should be noted that children talked about primary sources nearly four times as often as they mentioned secondary sources of light.

The nature of children's understanding of secondary sources was probed in more detail but only a minority of children were able to offer any explanation that approximated to a 'scientific' understanding of these sources. The data shown in Table 9.1 indicates that upper junior children were developing a fuller understanding of secondary sources of light. This meant that they were able to provide an explanation which recognised the true source of the light. However an examination of the data by splitting it by age, regardless of whether it was collected pre-intervention or post-intervention, shows evidence that this change is possibly occurring by experiential development.

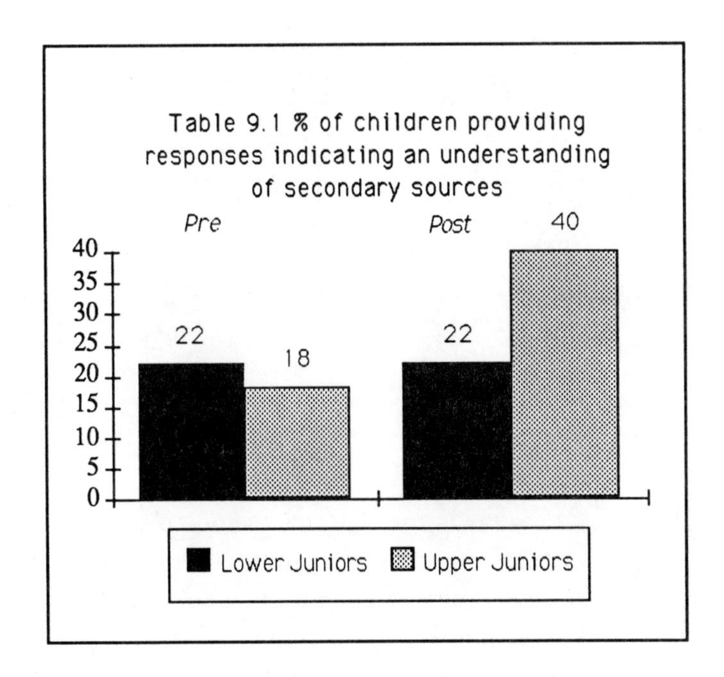

The most noticeable feature shown by the data is that there was very little change in children's understanding of sources of light as a result of this intervention. Table 9.2 shows the mean number of sources provided by children of both groups before and after the intervention.

**Table 9.2. Mean Number of Primary Sources indicated**

| | *Pre-Intervention* | | *Post-Intervention* | |
|---|---|---|---|---|
| | Lower Juniors | Upper Juniors | Lower Juniors | Upper Juniors |
| Mean number of primary sources shown per individual | 3.5 | 4.0 | 3.6 | 4.0 |

## *9.2 Children's representations of light:*

The representations used by children to show light were another prominent feature of children's work. Firstly, it was apparent that nearly all children's drawings would show light around sources represented by short lines (Table 9.3).

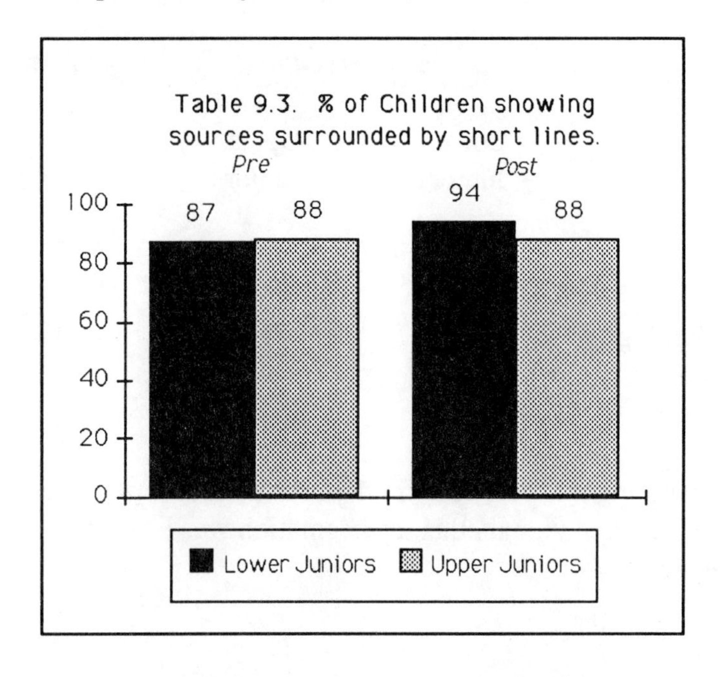


However the use of extensive lines e.g lines that linked source and the object or source and the eye, was limited prior to the intervention (Table 9.4). One notable effect of the intervention was the increase in the number of children who used this form of representation for light in their responses. Upper Junior children in particular made significantly greater use of lines with arrows to indicate a sense of direction.

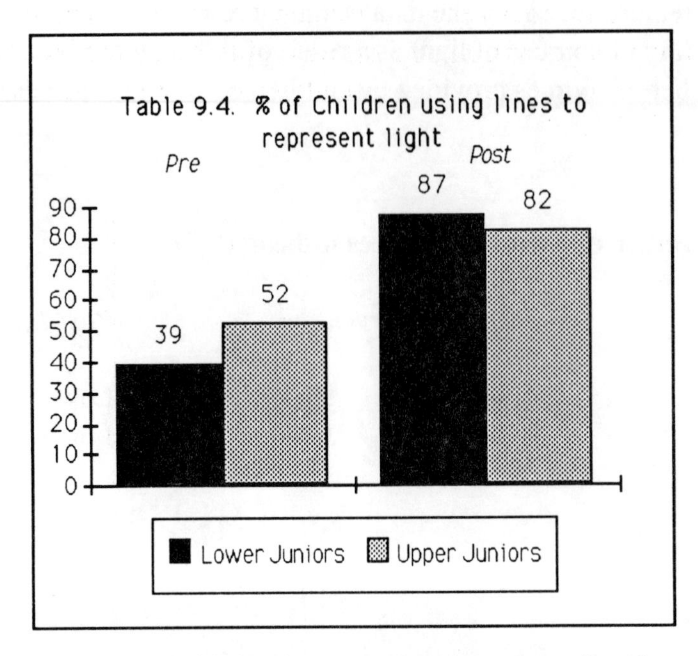


Another notable feature of children's representations were that those used by the upper junior children became more varied and context dependent. Significantly more children used dual representations of light in an elicitation of their understanding after the intervention. These data are shown in Table 9.5. A closer examination of the data shows that part of this is accounted for by an increase in the use of beams to represent light.

**Table 9.5. Percentage of children using more than one representation for light**

| | *Pre-Intervention* | | *Post-Intervention* | |
|---|---|---|---|---|
| | Lower Juniors | Upper Juniors | Lower Juniors | Upper Juniors |
| % of children providing dual representations | 26 | 27 | 48 | 61 |

Finally, it is worth noting that very few children showed no representation of light. For many lower junior children though, this representation was limited to 'simple' lines surrounding sources.

### *9.3 Children's responses to explain vision:*

Table 9.6 shows that many children were able to provide responses which indicated a link between eye and object. The majority of these responses incorporated a sense of direction to the link.

However, a sizeable proportion of lower junior children (35%) provided responses which showed no explanation for vision. A possible explanation is that vision is non-problematic for them and that 'we see with our eyes' was sufficient to account for the phenomena.

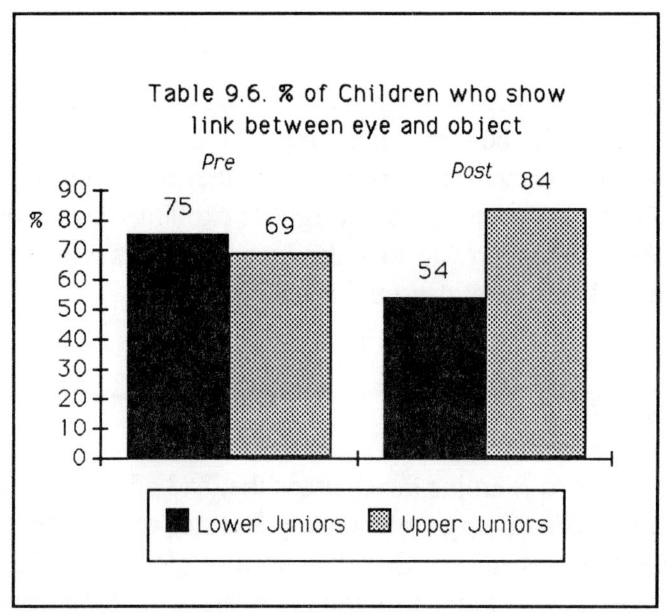


The most noticeable effect of the intervention on children's understanding was the increased use of dual links i.e eye-object and object-source by upper junior children (Table 9.7) coupled with a decreased use by lower junior children.

**Table 9.7 Percentage of responses using dual links to explain vision**

| | *Pre-Intervention* | | *Post-Intervention* | |
|---|---|---|---|---|
| | Lower Juniors | Upper Juniors | Lower Juniors | Upper Juniors |
| % of responses | 30 | 7 | 18 | 45 |

However, this was accompanied by an increased use of dual models reflecting a greater context dependence of the response. The implication would be that whilst such work was developing children's understanding, their understanding was in a process of assimilation which is fluid. The children lacked any strong sense of coherence about their ideas which would allow a consistent interpretation of a range of experience. Another consequence would be that such work was possibly more appropriate to upper junior children as there was little evidence that the idea of dual links was assimilable by younger children

For lower junior children the two significant effects of the intervention were an increase in the number of responses showing single links between object and eye, and an associated decrease in those responses which provided no explanation of vision. From a developmental perspective, this could be regarded as a stage necessary for the development of a fuller understanding and a positive feature of this work. However there were no data obtained in this study which can confirm or refute this suggestion.

***d. Context dependence:***

A strong feature of the data obtained from the children was the context dependent nature of many of the responses within the same elicitation. The data to support this are shown in Table 7.15. There was insufficient time to probe whether the children perceived any conflict in their use of dual models to represent light or explain vision. For upper juniors, the effect of the intervention was to make such responses more common and there is little evidence to clarify what caused this change.

***e. Change in Individual Children***

The data were also analysed to examine the changes that had occurred in individual children. This was done by identifying common key features of children's responses e.g whether they used lines or blobs and whether they used single links or dual links. Similar questions were used before and after the intervention and the responses analysed to establish whether a move towards features of a more scientific understanding had occurred; whether the child's response showed features indicating regression or whether there had been no change in their response. These changes are represented by the concept development maps shown in Tables 8.1 to 8.4. Summary figures for individual changes in children's representation of light and their explanations for vision are shown in table 9.8 and 9.9.

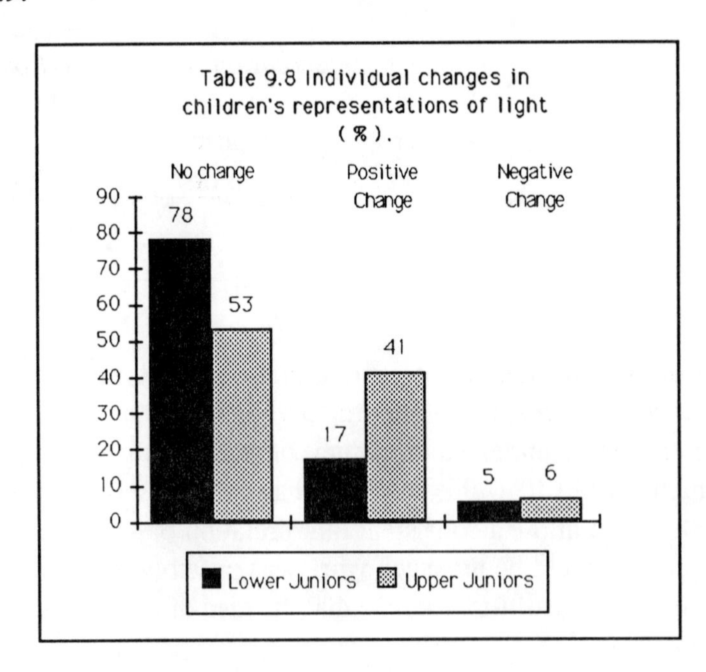


These show clearly that the predominant effect of the intervention has been positive in that children's responses show more features of a scientific understanding of light and that there was a greater change for upper junior children. However, the major feature is that for a large majority of children, there was no change in their response. This analysis supports that of the networks which show that the changes were predominantly for upper juniors who show more features of a scientific understanding.

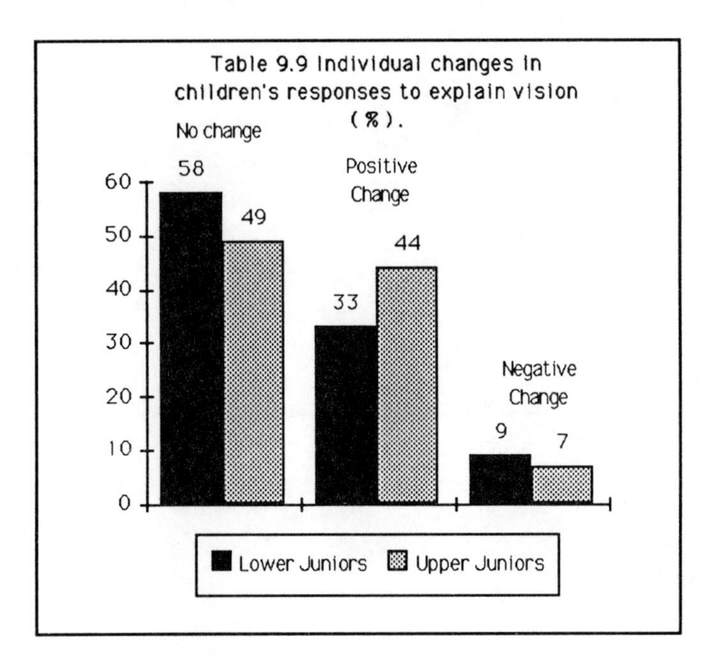


A closer analysis of the data shows that there was less development in the representations and responses to explain phenomena associated with secondary sources such as a book and a clock (Table 8.8 and 8.9). Very few children showed a positive change here and this would indicate that understanding how we see objects which are not primary sources of light was an area of conceptual difficulty for primary children.

# APPENDIX 1

## *Schools and teachers participating in the project*

### *ILEA*

Inspector for Science Education: John Wray

| *Schools* | *Headteacher* | *Teachers* |
|---|---|---|
| Ashmole Primary | Ms P.Turnbull | Mrs M. Hutchinson |
| Henry Fawcett Junior | Mr F.S. Curle | Ms R. Newlove<br>Mrs M. Robinson |
| Johanna Primary | Mr J.W. Hines | Mrs D. Carter<br>Ms R.Hines<br>Ms H. Ogbonna |
| Vauxhall Primary | Miss G. Brunt | Ms D. Gordon<br>Mr V. Hayes |
| Walnut Tree Walk | Mrs V. Phahle | Mrs Wai-choo Tsang |

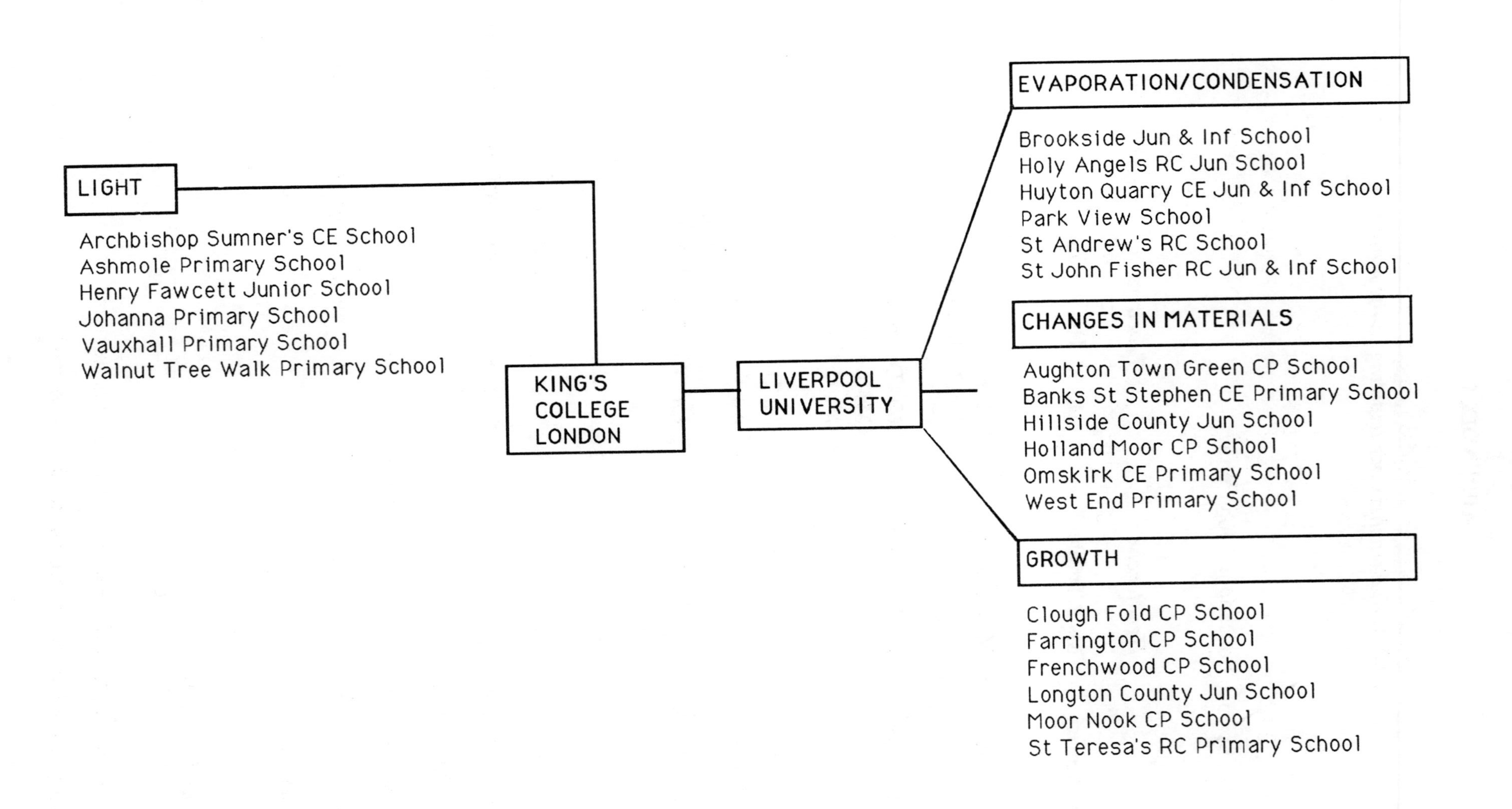

LIGHT
Archbishop Sumner's CE School
Ashmole Primary School
Henry Fawcett Junior School
Johanna Primary School
Vauxhall Primary School
Walnut Tree Walk Primary School
KING'S COLLEGE LONDON
LIVERPOOL UNIVERSITY
EVAPORATION/CONDENSATION
Brookside Jun & Inf School
Holy Angels RC Jun School
Huyton Quarry CE Jun & Inf School
Park View School
St Andrew's RC School
St John Fisher RC Jun & Inf School
CHANGES IN MATERIALS
Aughton Town Green CP School
Banks St Stephen CE Primary School
Hillside County Jun School
Holland Moor CP School
Omskirk CE Primary School
West End Primary School
GROWTH
Clough Fold CP School
Farrington CP School
Frenchwood CP School
Longton County Jun School
Moor Nook CP School
St Teresa's RC Primary School

## APPENDIX 2

### Questions used in initial phase of research.

*These questions were used in the first phase of the research to explore children's understanding of the topic and to evaluate which questions were of particular value in eliciting such understanding.*

***Light***

1. Where is light coming from at the moment?
2. How does light get here?
3. What happens to light at night?
4. What is the darkest place that you can remember?
5. Why was it so dark?
6. How do light reflectors work?
7. How can we find out which is the best reflector?
8. Can cats see in the dark?
9. Where is the light in this room?
10. Where does the light come from at night when you are watching TV?
11. Write down as many places as possible where light comes from.
12. Very young children think they have disappeared if they cover their head. Why do they believe this?
13. Look at the candle. What has to happen for you to see it?
14. How would you explain seeing to a younger child? Do a drawing to show what you mean.

***Mirrors and Torches***

1. Where are all the different places that mirrors are used?
2. Can you use a mirror to see this torch behind you?

3. What did you have to do to see the torch?
4. What do you see when you look in a mirror?
5. Would you be able to use a mirror in the dark?
6. What is a mirror-image?

*Colour and Light*

1. Close your eyes. Now close your eyes and cover them with your hands. Does it look different? If so, can you describe the difference?
2. Which colours are easiest to see
   a) in daylight?
   b) at night?
3. How could you test your answers to (2)?
4. Would we get the same answer for everybody?

*Seeing*

1. Can you see the light I have placed over there?
2. Why can some people see the light but others cannot?
3. Draw how you think we see the light.

*Light and Shadows*

1. Why does it go dark just before a storm?
2. What are shadows?
3. Where do you find shadows?
4. What are shadow puppets?

# APPENDIX 3

## Activities A - F

### *A. Places where light comes from*

EQUIPMENT REQUIRED Drawing Paper
Pencils

*Questions*

1. Where is light coming from at the moment?

2. How does light get here from the sun?

3. What happens to light at night?

4. Draw pictures of all the things that give off light.

---

### *B. Reflectors*

EQUIPMENT REQUIRED Plastic Bicycle Reflector
Torch
Drawing paper and pencils
Tape recorder

1. Switch on the torch and shine it on the reflector.

   How do you think reflectors work?

2. Would the reflector work in absolute darkness?

   Why not?

3. Do a drawing to show how the reflector works when the torch is shone on it.

*C. Torch and Mirror*

EQUIPMENT Torch for each pair of children
Mirror
Drawing paper and pencils

1. Activity: One child holds the torch which is switched on behind the child's head. The second child is seated and given a plane mirror. He/she is asked to use the mirror to see the light from the torch.

2. Do a drawing to show how you used the mirror to see the light from the torch behind you.

3. Show on the drawing how you think the light travels.

4. Is any light coming towards you?

5. How would you explain what is happening?

---

*D. Torch shining on paper*

EQUIPMENT Torch
Piece of plain paper
Drawing paper and pencil

1. Switch the torch on and shine it at the piece of card.
   What do you see on the card?

2. How does light get to the card?

3. What happens to the light at the card?

4. Do a drawing to show what is happening when the torch is shone on the card.

---

*E. Lighted Candle*

EQUIPMENT NEEDED Candle standing in sand tray
Matches
Drawing papers and pencils

Activity: Teacher lights the candle

1. Do a drawing to show how you see the light from the candle?
2. How far does light from the candle travel?
3. Could you see the candle burning from the other side of a big room?

   Why is this? Explain your answer.
4. How do you think you see the light from the candle?

---

*F. Seeing*

EQUIPMENT NEEDED

Matches
Book
Drawing of 2 pupils looking at clock
Drawing paper & pencils

1 (a) Look at the book. How would you explain to you younger brother or sister how we see the book?

(b) Do a drawing to show how we see the book.

2. Explain what happens to our sight if there is no light. ?

3. How does light help us to see?

4. Look at the drawing beneath which shows two children in a classroom

   Add to the drawing to show how you think the children see the clock.

***Additional Questions***

The following are two additional questions that were added to the activities to be used in the elicitation after the intervention.

1. Look at this diagram. It shows a box from on top with two holes, a mirror and a torch. Add to the picture to show where the light goes.

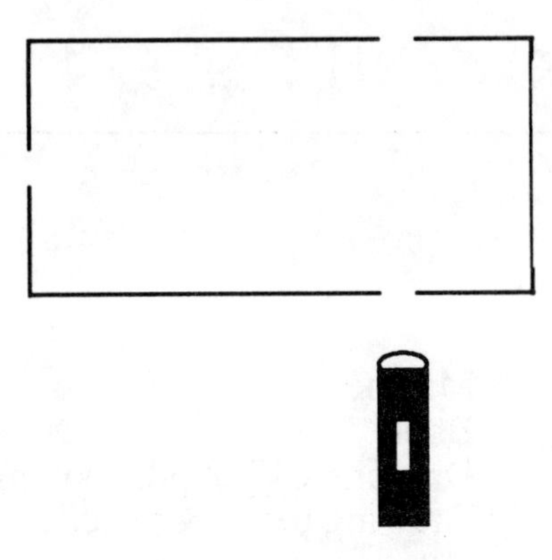

Now add to this picture to show where the light is here.

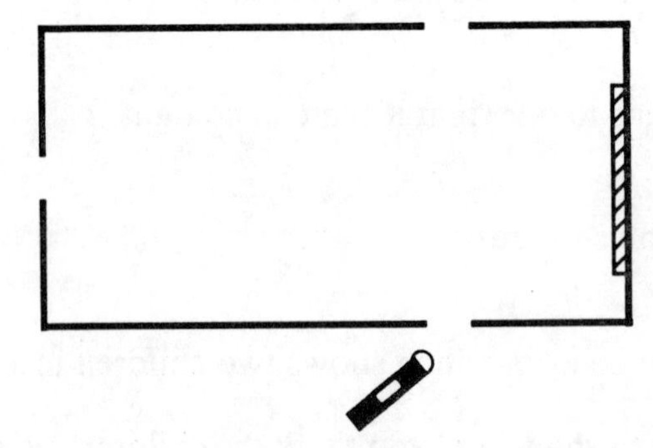

2. Light is all around us. Write three sentences about light. Try and include the word 'light' in your sentence.

# APPENDIX 4

## Experiences with Light

## Teachers' Notes

### Bouncing light around a Table

Equipment Needed

Torch (Fairly bright, powerful torch needed)
4 Plastic Mirrors
Piece of white card
Plasticene

Aims

a) To introduce children to the idea that light can be reflected off shiny objects.

b) To develop the idea that light is travelling from one object to another.

c) To develop a model of representing light in drawings and diagrams.

This exercise should be posed as a simple game with light for children. The object is to send light from the torch by bouncing from one mirror to another till it is returned to the eye of the first person. The diagram shows the normal arrangement for doing this. Children in groups of four, should be introduced to this as a problem/game which they are asked to hypothesise an answer to first i.e

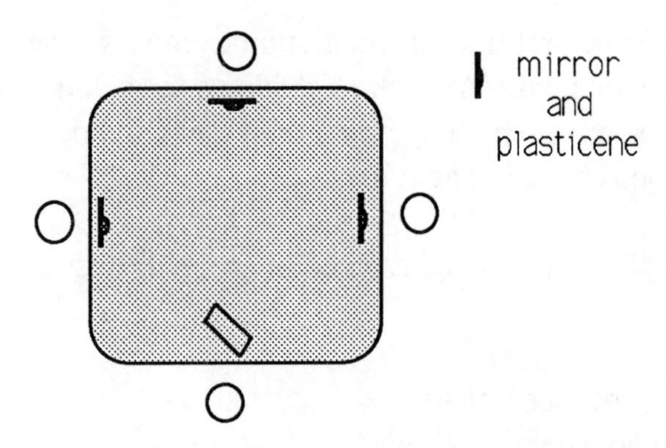


How could we bounce the light from this torch around the four sides of the table?

The activity can be structured by dividing it into four tasks:

Activity 1: Draw a diagram to show how you think light from the torch could be sent around the four sides of a table.

Activity 2: Using the mirrors, see if you can do this as a group.

Activity 3: Draw a diagram to show how you managed to do this task.

Activity 4: Imagine that you are a scientist, trying to find out a bit more about light. What would this activity have told you about light?

***Investigating Shadows***

EQUIPMENT NEEDED

Torch
Small Stick
Ruler
Scissors
Shoe Box
Cocktail Sticks
Plasticene
Pencil
Card
Paper
Sellotape

Aims:

a) To develop the idea that light is travelling from one object to another

b) To develop a model of representing light in drawings and diagrams

c) To provide an opportunity to examine the idea that shadows are formed by blockages of light.

d) To develop the idea that sharp shadows form because light travels in straight lines.

The shadow activities can be presented as prediction exercises. The children can be asked to guess or predict where different sized shadows form and then test their predictions. This allows them to challenge their own ideas and develop them. Children can work in pairs or groups for these activities .

Before the activity, the children can be involved in group discussion provoked by such questions as

What produces a shadow?
When do we get shadows?
Are shadows sharp or fuzzy?
Why are shadows sharp?

Children should have an opportunity to discuss these questions and record their ideas.

Discussion can be followed by the following tasks.

Activity 1: Draw where you think a shadow will form when a torch is shone on a pencil.

Activity 2: Try out this activity. Record your result.

Activity 3: Draw where you think a torch should be held to obtain

a) A shadow which is larger than a pencil.
b) A shadow which is smaller than a pencil.

Activity 4: Try out this activity. Record what you found out as drawing.

Activity 5: Where can you place a torch so that it shines on a stick and produces no shadow?

*The Light Boxes Activity*

EQUIPMENT NEEDED

1 Shoe Box
Mirror
Torch

Aims:

a) To provide children an opportunity to explore how light travels.

b) To develop a model that light travels and travels in straight lines.

c) To see that light can be bounced off mirrors.

d) To observe that light cannot be seen travelling from one place to another.

The light box has two small holes on opposite sides, a viewing slot and a small mirror taped to the inside back wall.

First ask the children to guess what they think will happen to the light when the torch is shone into one of the small holes. A worksheet is provided with suggestions for activities which can be used here.

The activity can then be done with the children working in pairs - one child looks through the viewing slot, preferably in a dark or shaded room. Another child shines the torch into one of the holes, first directly across the box, then at an angle onto the mirror. The children then swop roles and can be asked to discuss what they saw, where they thought the light was in the box and whether they had changed their minds from their original guesses. They can repeat this activity, looking into different holes until they are ready to complete the second activity. In this they are asked to complete drawings of the inside of the boxes showing where the light is, and whether they have changed their minds now that they have used the boxes with the torches.